Physical Properties of Hydrocarbons

VOLUME 1

SECOND EDITION

Gulf Publishing Company
Houston, London, Paris, Zurich, Tokyo

Physical Properties of Hydrocarbons

VOLUME 1

SECOND EDITION

R. W. GALLANT

CARL L. YAWS

Physical Properties of Hydrocarbons
Volume 1/Second Edition

Copyright © 1968, 1992 by Gulf Publishing Company, Houston, Texas. All rights reserved. Printed in the United States of America. This book, or parts thereof, may not be reproduced in any form without permission of the publishers.

Gulf Publishing Company
Book Division
P.O. Box 2608   Houston, Texas 77252-2608

10 9 8 7 6 5 4 3 2 1

Printed on Acid-Free Paper (∞)

Library of Congress Cataloging-in-Publication Data
(Revised for vol. 1)
Gallant, Robert W.
    Physical properties of hydrocarbons.
    Vol. 1 by Robert W. Gallant, Carl L. Yaws.
    Includes bibliographical references and indexes.
    1. Hydrocarbons—Tables.   I. Railey, Jay M.   II. Yaws, Carl L.
III. Title.
QD305.H5G28 1984   547′.01   83-22604
ISBN 0-88415-067-4 (v. 1)   ISBN 0-87201-690-0 (v. 2)

CONTRIBUTORS

Robert W. Gallant	Vice President, Texas Operations, Dow Chemical Company, Freeport, Texas 77641, U S A
Jack R. Hopper	Professor and Chair, Chemical Engineering Department, Lamar University, Beaumont, Texas 77710, U S A
Xiang Pan	Process Engineer, Jacobs Engineering Group Inc., Houston, Texas 77052, U S A
Duane G. Piper, Jr.	Senior Engineer, E. I. Du Pont de Nemours & Co. Inc., Beaumont Works, Beaumont, Texas 77704, U S A
Carl L. Yaws	Professor, Chemical Engineering Department, Lamar University, Beaumont, Texas 77710, U S A

ACKNOWLEDGMENTS

    Many colleagues and students have made contributions and helpful comments over the years. The author is grateful to each: Jack R. Hopper, Joe W. Miller, Jr., C. S. Fang, K. Y. Li, Keith C. Hansen, Daniel H. Chen, P. Y. Chiang, H. C. Yang and Xiang Pan.
    The author wishes to acknowledge his special appreciation and dedication to his wife (Annette) and children (Kent, Rebecca and Sarah) for their help and understanding of the week-end and night labors.
    The author wishes to acknowledge that the Gulf Coast Hazardous Substance Research Center provided partial support to this work. A special note of thanks to William C. Cawley (Director) is also acknowledged.

Carl L. Yaws

CONTENTS

Chapter		Page
1.	C_1 TO C_4 NORMAL ALKANES	1
2.	C_2 TO C_4 MONOOLEFINS	14
3.	C_2 TO C_4 ALKYNES	24
4.	C_2 TO C_4 DIOLEFINS	34
5.	CHLORINATED METHANES	44
6.	CHLORINATED ETHYLENES	54
7.	CHLORINATED ALIPHATICS	64
8.	PRIMARY ALCOHOLS	74
9.	C_3 TO C_4 ALCOHOLS	84
10.	MISCELLANEOUS ALCOHOLS	94
11.	C_2 TO C_4 OXIDES	104
12.	ETHYLENE GLYCOLS	114
13.	PROPYLENE GLYCOLS AND GLYCERINE	135
14.	C_5 TO C_8 ALKANES	155
15.	C_5 TO C_8 ALKENES	165
16.	C_4 TO C_5 BRANCHED HYDROCARBONS	175
17.	C_6 TO C_8 BRANCHED HYDROCARBONS	185
18.	CHLORINATED C_2 COMPOUNDS	195
19.	HALOGENATED METHANES	205
20.	HALOGENATED HYDROCARBONS	215
21.	FLUORINATED HYDROCARBONS	225
22.	BROMINATED HYDROCARBONS	235
	REFERENCES	245

Appendix		
A.	CONVERSION TABLES	253
B.	COMPOUND INDEX BY FORMULA	254
C.	COMPOUND INDEX BY NAME	256

PREFACE

The objective of PHYSICAL PROPERTIES OF HYDROCARBONS, Volume 1 (2nd edition), is to provide the working engineer with all the essential data to design and run production facilities. The data are presented in graphs covering a wide temperature range to enable the engineer to quickly determine the information he needs at the desired temperature.

The literature has been carefully searched for experimental data. When several sources give different values, the authors made a judgment as to the most reliable source. To enable the engineer to go back to the original article for additional information, the sources of all data are documented.

For many compounds, experimental results are available only over a narrow temperature span. In these cases, estimation methods have been used to extend the data over a wider temperature range. When estimation methods are used, the method and expected accuracy are explained. Thus, the user is aware of the reliability of the graphical values. Reference to the original work is provided for those wishing to study the method further.

The physical properties normally needed in design and production are vapor pressure, heat of vaporization, density, surface tension, heat capacity and thermal conductivity. For chemical reactions, enthalpy of formation and Gibb's free energy of formation are helpful. Also, the boiling point, freezing point, molecular weight, critical properties, lower explosion limit in air and solubility in water are tabulated for each compound. This latter property data are helpful in safety and environmental engineering.

In this 2nd edition, special attention is paid to improving the accuracy of estimation techniques. Improved methods of extending data and new experimental data are included.

The SI and metric units are used for all properties except vapor pressure, both because these units are becoming increasingly used in production plants and because conversion is generally easier. Each graph displays a conversion factor providing English units. The temperature scale in all graphs is Centigrade.

Robert W. Gallant
Freeport, Texas

Carl L. Yaws
Beaumont, Texas

DISCLAIMER

This handbook presents a variety of thermodynamic and physical property data. It is incumbent upon the user to exercise judgment in the use of the data. The authors do not provide any guarantee, express or implied, with regard to the general or specific applicability of the data, the range of errors that may be associated with any of the data, or the appropriateness of using any of the data in any subsequent calculation, design or decision process. The authors accept no responsibility for damages, if any, suffered by any reader or user of this handbook as a result of decisions made or actions taken on information contained herein.

Chapter 1

C₁ TO C₄ NORMAL ALKANES

Robert W. Gallant, Carl L. Yaws and Xiang Pan

PHYSICAL PROPERTIES - Table 1-1

Physical, thermodynamic and transport property data from the literature (1-74) are given in Table 1-1. Results from the DIPPR (Design Institute for Physical Property Research) project (5) and recent data compilations by Yaws and co-workers (44-55) were consulted extensively in preparing the tabulation.

The critical constants (temperature, pressure, volume and compressibility factor) have been determined experimentally and are available (1-7). Additional property data such as acentric factor, enthalpy of formation, lower explosion limit in air and solubility in water are also available (8-74). The property data in the top and middle parts of the tabulation are helpful in process engineering. The property data in the lower part of the tabulation are helpful in safety and environmental engineering.

VAPOR PRESSURE - Figure 1-1

Results from the DIPPR project (5) were selected. These results are applicable for vapor pressure from very low temperatures to the critical point. Correlation of vapor pressure as a function of temperature was accomplished using the equation:

$$\ln P = A + B/T + C \ln T + D T^E \tag{1-1}$$

where
P = vapor pressure
T = temperature
A, B, C, D and E = correlation constants

Results from this equation (Antoine-type with extended terms) are in favorable agreement with experimental data. Errors are about 1-5% or less in most cases.

HEAT OF VAPORIZATION - Figure 1-2

Results from the data compilation of Yaws and co-workers (44,52) were selected. Data for heat of vaporization were correlated using the Watson equation:

$$\Delta H_{vap} = A (1 - T_r)^{0.38} \tag{1-2}$$

where
ΔH_{vap} = heat of vaporization
T_r = reduced temperature, T/T_C
A = correlation constant

Reliability of results is good with errors of about 1-5% or less.

For the DIPPR project (5), data for heat of vaporization were correlated using a modified Watson equation with extended terms in the exponent:

$$\Delta H_{vap} = A (1 - T_r)^{[B + C T_r + D T_r^2 + E T_r^3]} \tag{1-2a}$$

where
ΔH_{vap} = heat of vaporization
T_r = reduced temperature, T/T_C
A, B, C, D and E = correlation constants

Both of the above equations give good results for heat of vaporization. For these alkanes, a comparison of values from Equations (1-2) and (1-2a) provided essentially equivalent results.

LIQUID DENSITY - Figure 1-3

Results from the data compilation of Yaws and co-workers (44,54) were selected. A modified Rackett equation was used for correlation of the data:

$$\rho = A B^{-(1 - T_r)^{2/7}} \tag{1-3}$$

where
ρ = saturated liquid density
T_r = reduced temperature, T/T_C
A and B = correlation constants

Results from the correlation are in favorable agreement with data. Deviations are less than 1-2% in most cases.

For the DIPPR project (5), a slightly different Rackett equation was used for correlation of the data:

$$\rho = A B^{-(1 - T/C)^D} \tag{1-3a}$$

where
ρ = saturated liquid density
T = temperature
A, B, C and D = correlation constants

Both of the above equations give good results. For these alkanes, a comparison of values from Equations (1-3) and (1-3a) provided essentially equivalent results for liquid density.

SURFACE TENSION - Figure 1-4

Results from the data compilation of Yaws and co-workers (44,55) were selected. Using data from the literature, surface tension over the full liquid range was achieved by the modified Othmer equation:

$$\sigma = A (1 - T_r)^{11/9} \tag{1-4}$$

where σ = surface tension
T_r = reduced temperature, T/T_C
A = correlation constant

Accuracy is good with errors being about 1-10% or less in most cases.

For the DIPPR project (5), surface tension was correlated by a slightly different Othmer equation:

$$\sigma = A (1 - T_r)^{[B + C T_r + D T_r^2 + E T_r^3]} \tag{1-4a}$$

where σ = surface tension
T_r = reduced temperature, T/T_C
A, B, C, D and E = correlation constants

Both of the above equations give good results for surface tension. A comparison of values from Equations (1-4) and (1-4a) with experimental data (37) provided essentially equivalent results for methane and butane. For ethane and propane, better agreement of correlation and experimental data was achieved with Equation (1-4).

HEAT CAPACITY - Figures 1-5 and 1-6

Results from the data compilation of Yaws and co-workers (44) were selected for heat capacity of ideal gas. Correlation of data was accomplished using a series expansion in temperature:

$$C_p = A + B T + C T^2 + D T^3 \tag{1-5}$$

where C_p = heat capacity of ideal gas
T = temperature
A, B, C and D = correlation constants

Results are in favorable agreement with data. Errors are about 1% or less in most cases.

For the DIPPR project (5), a different equation was used for correlation of the data for heat capacity of ideal gas:

$$C_p = A + B \exp[-C/T^D] \tag{1-5a}$$

where C_p = heat capacity of ideal gas
T = temperature
A, B, C and D = correlation constants

Both of the above equations give good results. For these alkanes, a comparison of values from Equations (1-5) and (1-5a) provided essentially equivalent results.

Results from the DIPPR project (5) were selected for heat capacity of liquid. Data were correlated with a series expansion in temperature:

$$C_p = A + B T + C T^2 + D T^3 + E T^4 \tag{1-6}$$

where C_p = heat capacity of liquid
T = temperature
A, B, C, D and E = correlation constants

Results are in favorable agreement with data. Errors are about 5% or less using the correlation.

VISCOSITY - Figures 1-7 and 1-8

Results from the DIPPR project (5) were selected for viscosity of gas. Data were correlated using the equation:

$$\eta_{gas} = \frac{A\, T^B}{1 + C/T + D/T^2} \qquad (1\text{-}7)$$

where η_{gas} = viscosity of gas
T = temperature
A, B, C and D = correlation constants

Results are in favorable agreement with data. Errors are about 1-10% or less in most cases.

Results from the DIPPR project (5) were selected for viscosity of liquid. Data were correlated using the de Guzman - Andrade equation with extended terms:

$$\ln \eta_{liq} = A + B/T + C \ln T + D\, T^E \qquad (1\text{-}8)$$

where η_{liq} = viscosity of liquid
T = temperature
A, B, C, D and E = correlation constants

Correlation results and data are in favorable agreement with errors being about 1-5% or less.

THERMAL CONDUCTIVITY - Figures 1-9 and 1-10

Results from the DIPPR project (5) were selected for thermal conductivity of gas. Data were correlated using the equation:

$$\lambda_{gas} = \frac{A\, T^B}{1 + C/T + D/T^2} \qquad (1\text{-}9)$$

where λ_{gas} = thermal conductivity of gas
T = temperature
A, B, C and D = correlation constants

Reliability of results is good with errors of about 1-10% or less in most cases.

Results from the DIPPR project (5) were selected for thermal conductivity of liquid. Data were correlated using the equation:

$$\lambda_{liq} = A + B\,T + C\,T^2 + D\,T^3 + E\,T^4 \qquad (1\text{-}10)$$

where
λ_{liq} = thermal conductivity of liquid
T = temperature
A, B, C, D and E = correlation constants

Results are in favorable agreement with data. Errors are about 1-10% or less in most cases.

ENTHALPY OF FORMATION - Figure 1-11

Results from the data compilation of Yaws and co-workers (44,45) were selected. Data for enthalpy of formation of the ideal gas is a series expansion in temperature:

$$\Delta H_f = A + B\,T + C\,T^2 \qquad (1\text{-}11)$$

where
ΔH_f = enthalpy of formation of ideal gas
T = temperature
A, B and C = correlation constants

Results from the correlation are in favorable agreement with data.

GIBB'S FREE ENERGY OF FORMATION - Figure 1-12

Results from the data compilation of Yaws and co-workers (44,46) were selected. Data for Gibb's free energy of formation of the ideal gas is a series expansion in temperature:

$$\Delta G_f = A + B\,T + C\,T^2 \qquad (1\text{-}12)$$

where
ΔH_f = Gibb's free energy of formation of the ideal gas
T = temperature
A, B and C = correlation constants

Results from the correlation are in favorable agreement with data.

Table 1-1 Physical Properties

	Methane	Ethane	Propane	Butane
1. Name	Methane	Ethane	Propane	Butane
2. Formula	CH_4	C_2H_6	C_3H_8	C_4H_{10}
3. Molecular Weight, g/mol	16.043	30.069	44.096	58.123
4. Critical Temperature, K	190.58	305.42	369.82	425.18
5. Critical Pressure, bar	46.043	48.801	42.492	37.969
6. Critical Volume, ml/mol	99.25	147.92	202.88	255.00
7. Critical Compressibility Factor	0.288	0.284	0.280	0.274
8. Acentric Factor	0.0108	0.0990	0.1517	0.1931
9. Melting Point, K	90.67	90.35	85.46	134.79
10. Boiling Point @ 1 atm, K	111.66	184.55	231.11	272.65
11. Heat of Vaporization @ BP, kJ/kg	509.73	489.19	425.74	385.61
12. Density of Liquid @ 25 C, g/ml	------	0.321	0.493	0.573
13. Surface Tension @ 25 C, dynes/cm	------	0.514	5.67	11.89
14. Heat Capacity of Gas @ 25 C, J/g K	2.22	1.75	1.67	1.68
15. Heat Capacity of Liquid @ 25 C, J/g K	------	------	2.68	2.44
16. Viscosity of Gas @ 25 C, micropoise	111.40	93.76	82.76	75.43
17. Viscosity of Liquid @ 25 C, centipoise	------	------	0.106	0.162
18. Thermal Conductivity of Gas @ 25 C, W/m K	0.0337	0.0214	0.0181	0.0162
19. Thermal Conductivity of Liquid @ 25 C, W/m K	------	0.0762	0.0950	------
20. Enthalpy of Formation of Gas @ 25 C, kJ/mol	-74.83	-84.70	-103.94	-126.26
21. Gibbs Free Energy of Formation of Gas @ 25 C, kJ/mol	-50.96	-33.14	-23.77	-17.50
22. Flash Point, K	------	------	------	------
23. Autoignition Temperature, K	810.37	788.15	723.15	678.15
24. Lower Explosion Limit in Air, vol %	5.0	3.0	2.1	1.8
25. Upper Explosion Limit in Air, vol %	15.0	12.5	9.5	8.5
26. Solubility in Water @ 25 C, ppm(wt)	24.40	60.40	62.40	61.40

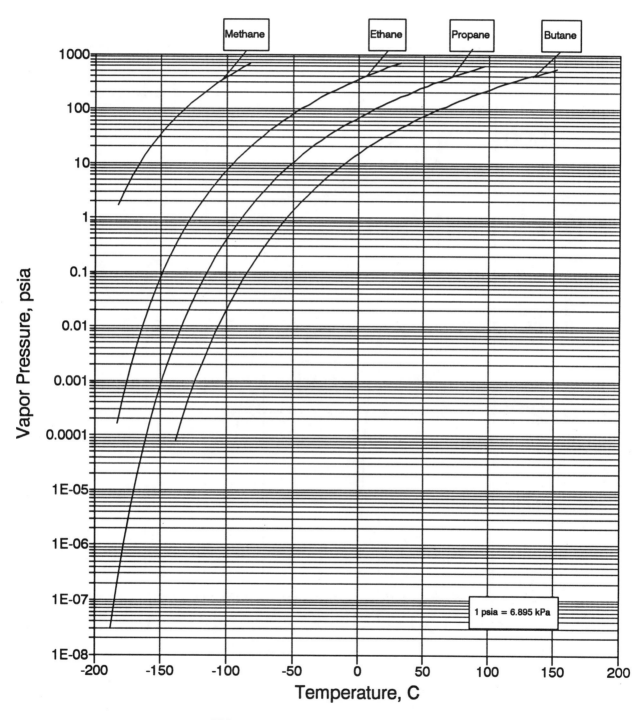

Figure 1-1 Vapor Pressure

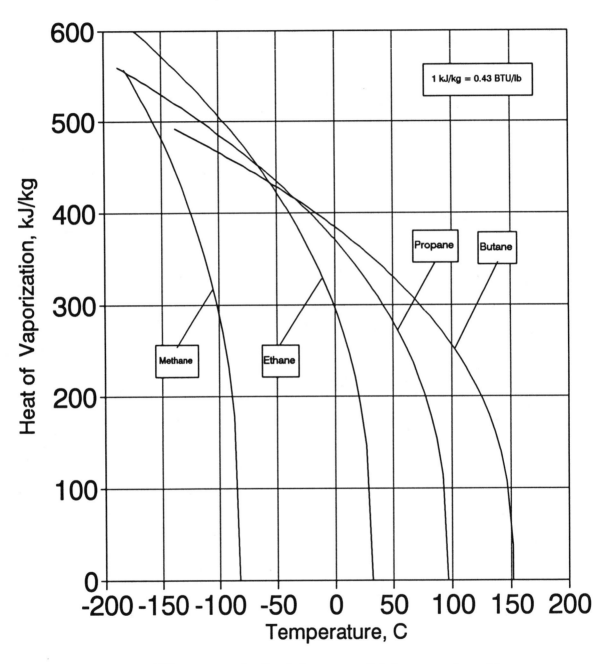

Figure 1-2 Heat of Vaporization

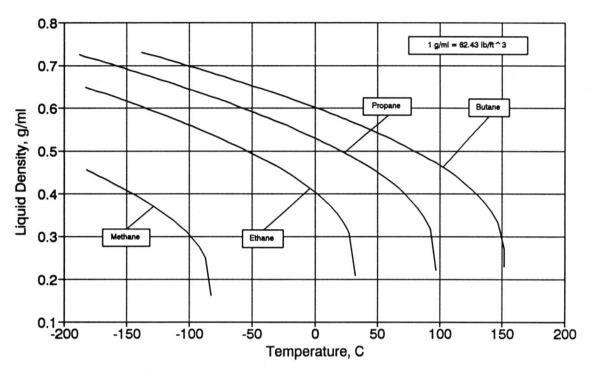

Figure 1-3 Liquid Density

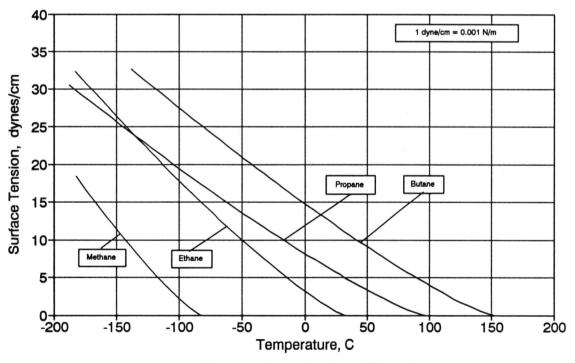

Figure 1-4 Surface Tension

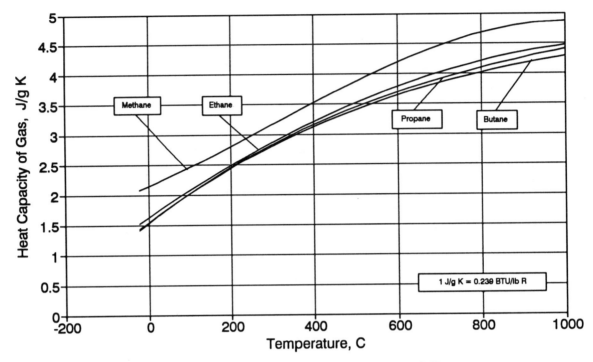

Figure 1-5 Heat Capacity of Gas

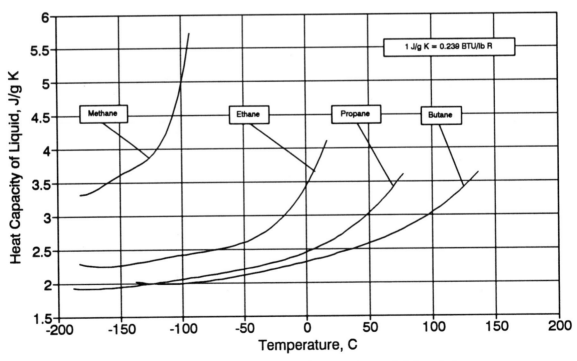

Figure 1-6 Heat Capacity of Liquid

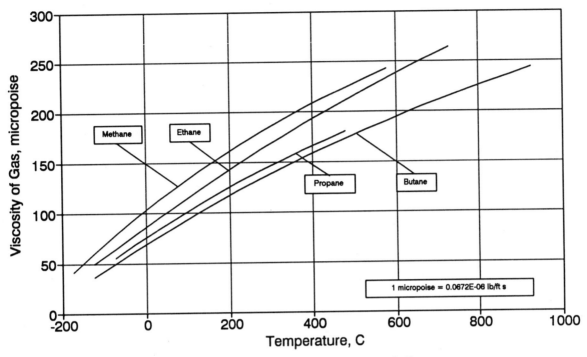

Figure 1-7 Viscosity of Gas

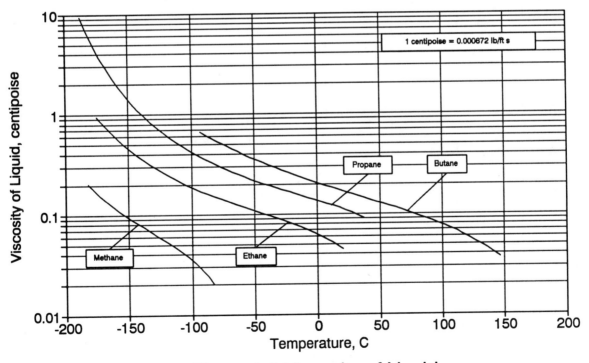

Figure 1-8 Viscosity of Liquid

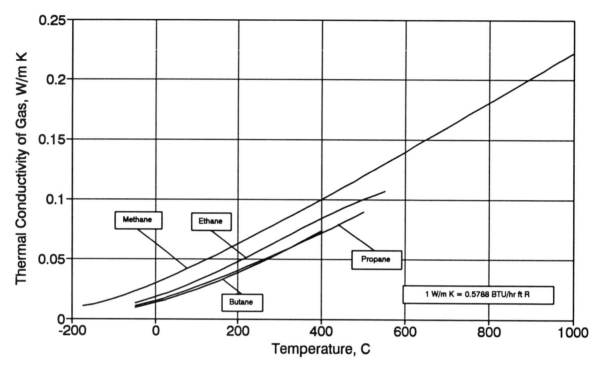

Figure 1-9 Thermal Conductivity of Gas

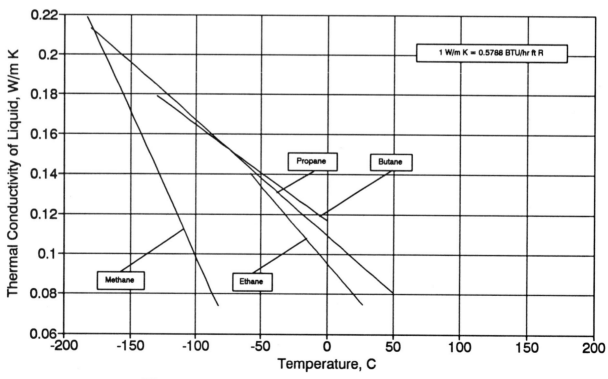

Figure 1-10 Thermal Conductivity of Liquid

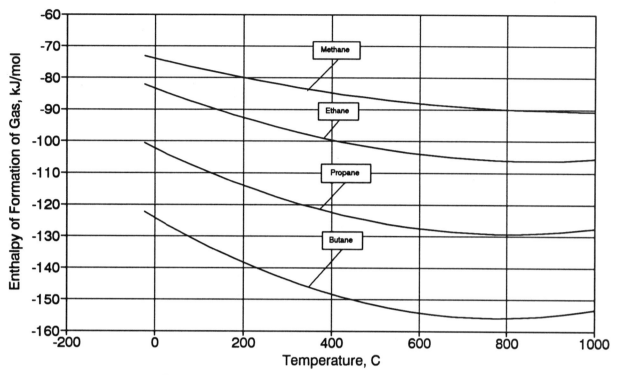

Figure 1-11 Enthalpy of Formation of Gas

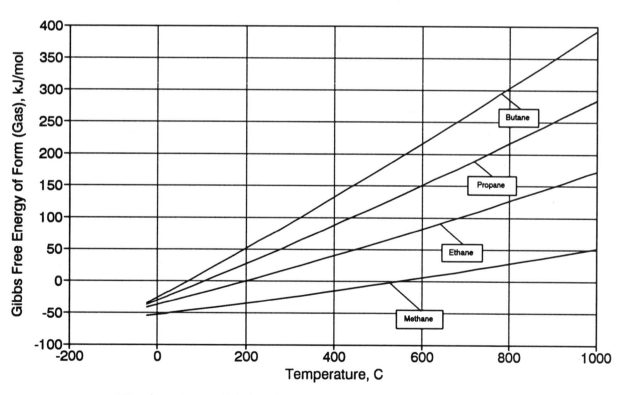

Figure 1-12 Gibbs Free Energy of Formation of Gas

Chapter 2

C_2 TO C_4 MONOOLEFINS

Robert W. Gallant, Carl L. Yaws and Jack R. Hopper

PHYSICAL PROPERTIES - Table 2-1

Property data from the literature (1-55,58,75-83) are given in Table 2-1. The critical constants (temperature, pressure, volume and compressibility factor) have been determined experimentally and are available (1-7). Additional property data such as acentric factor, enthalpy of formation, lower explosion limit in air and solubility in water are also available. The DIPPR (Design Institute for Physical Property Research) project (5) and recent data compilations by Yaws and co-workers (44-55) were consulted extensively in preparing the tabulation.

VAPOR PRESSURE - Figure 2-1

Results from the DIPPR project (5) were selected for vapor pressure from very low temperatures to the critical point. Correlation of data for vapor pressure as a function of temperature was accomplished using Equation (1-1). Results from this equation (Antoine-type with extended terms) are in favorable agreement with experimental data. Errors are about 1-5% or less in most cases.

HEAT OF VAPORIZATION - Figure 2-2

The data compilation of Yaws and co-workers (52) was selected for heat of vaporization for temperatures ranging from melting point to critical point. The Watson equation, Equation (1-2), was used for correlation of the data as a function of temperature. Reliability of results is good with errors of about 1-5% or less.

LIQUID DENSITY - Figure 2-3

Results from the data compilation of Yaws and co-workers (54) were selected for liquid density from low temperatures at the melting point to higher temperatures up to the critical point. A modified Rackett equation, Equation (1-3), was used for correlation of the data as a function of temperature. Results from the correlation are in favorable agreement with data. Deviations are less than 1-2% in most cases.

SURFACE TENSION - Figure 2-4

The data compilation of Yaws and co-workers (55) was selected for surface tension for temperatures from melting point to critical point. Using data from the literature, correlation for surface tension as a function of temperature over the full liquid range was achieved by the modified Othmer equation, Equation (1-4). Accuracy is good with errors being about 1-10% or less in most cases.

HEAT CAPACITY - Figures 2-5 and 2-6

Results from the data compilation of Yaws and co-workers (44) were selected for heat capacity of ideal gas. Correlation of data was accomplished using a series expansion in temperature, Equation (1-5). Results are in favorable agreement with data. Errors are about 1% or less in most cases.

Results from the DIPPR project (5) were selected for heat capacity of liquid. The coverage applies to temperatures from below the boiling point to temperatures above the boiling point for most of the compounds. Data were correlated with a series expansion in temperature, Equation (1-6). Results are in favorable agreement with data. Errors are about 5% or less using the correlation.

VISCOSITY - Figures 2-7 and 2-8

The DIPPR project (5) was selected for viscosity of gas. Data for gas viscosity as a function of temperature were correlated using Equation (1-7). Results are in favorable agreement with data. Errors are about 1-10% or less in most cases.

The DIPPR project (5) was also selected for viscosity of liquid. Temperatures from below the boiling point to temperatures above the boiling point are covered for most of the compounds. Data for liquid viscosity as a function of temperature were correlated using the de Guzman - Andrade equation with extended terms, Equation (1-8). Correlation results and data are in favorable agreement with errors being about 1-5% or less.

THERMAL CONDUCTIVITY - Figures 2-9 and 2-10

Results from the DIPPR project (5) were selected for thermal conductivity of gas. Data for gas thermal conductivity as a function of temperature were correlated using the Equation (1-9). Reliability of results is good with errors of about 1-10% or less in most cases.

Results from the DIPPR project (5) were selected for thermal conductivity of liquid. The coverage applies to temperatures from below the boiling point to temperatures above the boiling point for most of the compounds. Data for liquid thermal conductivity as a function of temperature were correlated using a series expansion in temperature, Equation (1-10). Results are in favorable agreement with data. Errors are about 1-10% or less in most cases.

ENTHALPY OF FORMATION - Figure 2-11

The data compilation of Yaws and co-workers (44,45) was selected for enthalpy of formation of ideal gas. Data for enthalpy of formation of the ideal gas is a series expansion in temperature, Equation (1-11). Results from the correlation are in favorable agreement with data.

GIBB'S FREE ENERGY OF FORMATION - Figure 2-12

Results from the data compilation of Yaws and co-workers (44,46) were selected for Gibb's free energy of formation of ideal gas. Data for Gibb's free energy of formation of the ideal gas is a series expansion in temperature, Equation (1-12). Correlation results are in favorable agreement with data.

Table 2-1 Physical Properties

	Ethylene	Propylene	1 Butene	Trans 2 Butene	Cis 2 Butene
1. Name	Ethylene	Propylene	1 Butene	Trans 2 Butene	Cis 2 Butene
2. Formula	C2H4	C3H6	C4H8	C4H8	C4H8
3. Molecular Weight, g/mol	28.054	42.080	56.107	56.107	56.107
4. Critical Temperature, K	282.36	364.76	419.59	428.63	435.58
5. Critical Pressure, bar	50.318	46.126	40.196	41.024	42.058
6. Critical Volume, ml/mol	129.07	181.00	239.93	238.18	233.98
7. Critical Compressibility Factor	0.277	0.275	0.276	0.274	0.272
8. Acentric Factor	0.0852	0.1424	0.1867	0.2182	0.2030
9. Melting Point, K	104.01	87.90	87.80	167.62	134.23
10. Boiling Point @ 1 atm, K	169.47	225.43	266.90	274.03	276.87
11. Heat of Vaporization @ BP, kJ/kg	482.67	437.82	390.61	405.57	416.15
12. Density of Liquid @ 25 C, g/ml	------	0.507	0.589	0.598	0.617
13. Surface Tension @ 25 C, dynes/cm	------	6.88	11.92	12.93	12.96
14. Heat Capacity of Gas @ 25 C, J/g K	1.55	1.52	1.53	1.56	1.40
15. Heat Capacity of Liquid @ 25 C, J/g K	------	2.45	2.28	------	2.28
16. Viscosity of Gas @ 25 C, micropoise	102.07	86.34	77.56	76.29	78.22
17. Viscosity of Liquid @ 25 C, centipoise	------	------	------	0.179	0.179
18. Thermal Conductivity of Gas @ 25 C, W/m K	0.0207	0.0173	0.0151	0.0144	0.0135
19. Thermal Conductivity of Liquid @ 25 C, W/m K	------	0.1006	------	------	------
20. Enthalpy of Formation of Gas @ 25 C, kJ/mol	52.30	20.42	-0.13	-11.17	-6.99
21. Gibbs Free Energy of Formation of Gas @ 25 C, kJ/mol	68.12	62.72	71.30	62.97	65.86
22. Flash Point, K	------	165.37	------	------	------
23. Autoignition Temperature, K	723.15	728.15	657.04	597.04	598.15
24. Lower Explosion Limit in Air, vol %	2.7	2.0	1.6	1.8	1.6
25. Upper Explosion Limit in Air, vol %	36.0	11.0	9.3	9.7	9.7
26. Solubility in Water @ 25 C, ppm(wt)	131.0	200.0	222.0	------	------

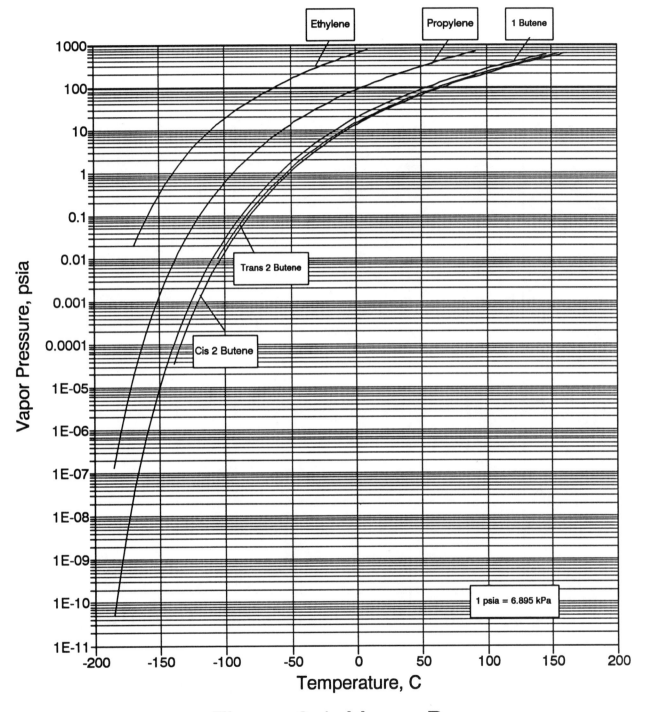

Figure 2-1 Vapor Pressure

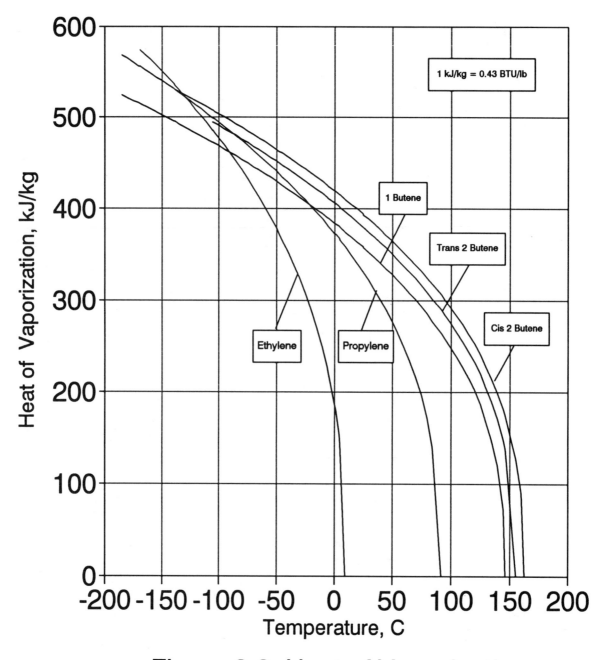

Figure 2-2 Heat of Vaporization

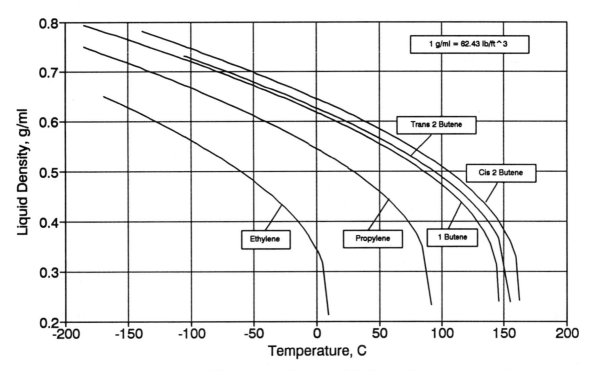

Figure 2-3 Liquid Density

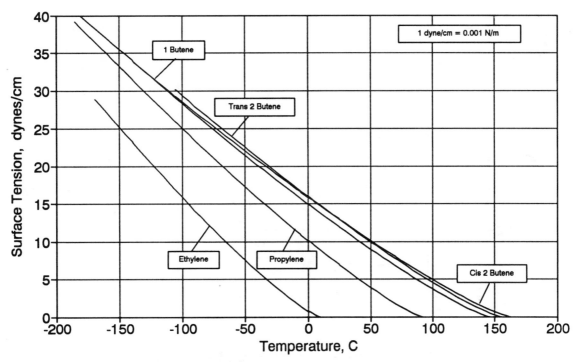

Figure 2-4 Surface Tension

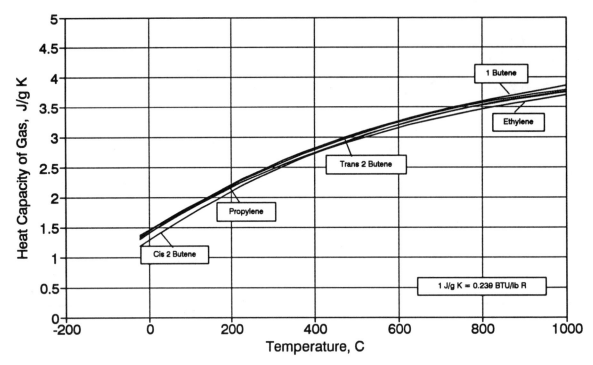

Figure 2-5 Heat Capacity of Gas

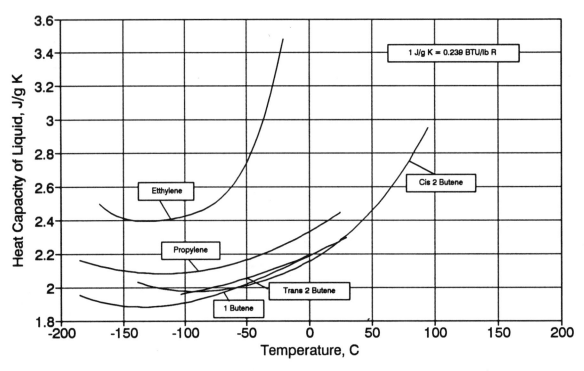

Figure 2-6 Heat Capacity of Liquid

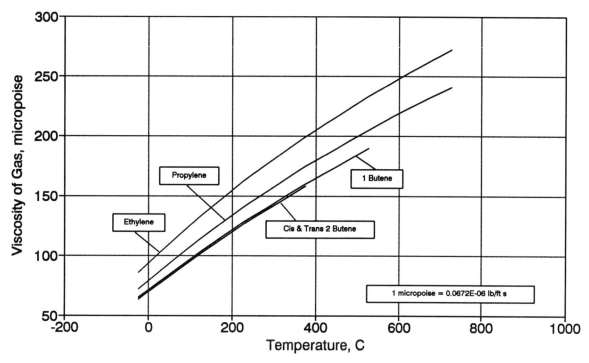

Figure 2-7 Viscosity of Gas

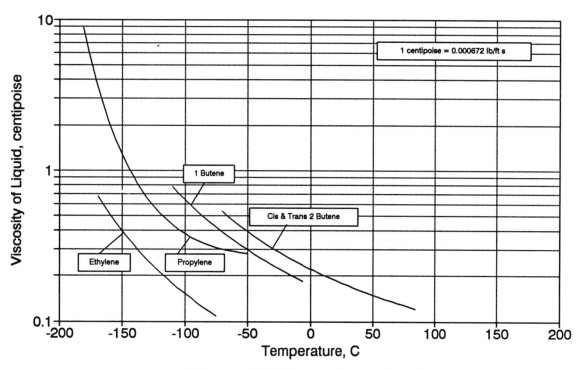

Figure 2-8 Viscosity of Liquid

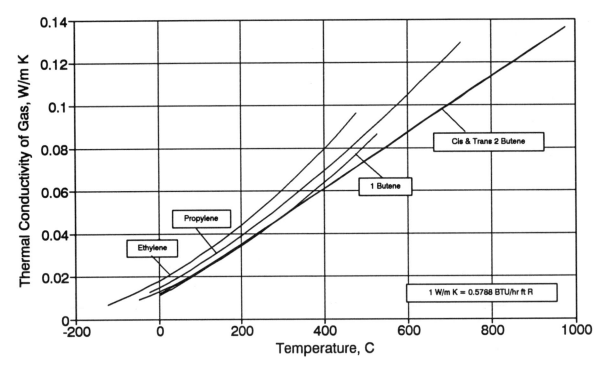

Figure 2-9 Thermal Conductivity of Gas

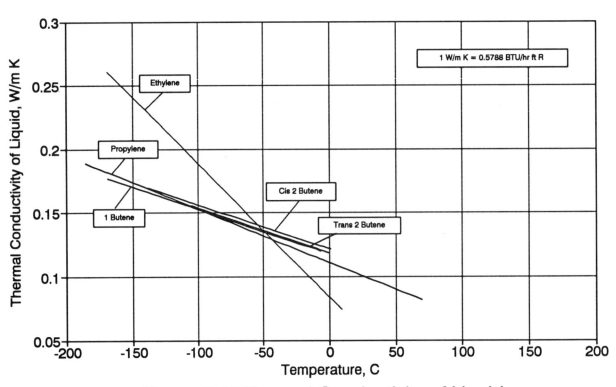

Figure 2-10 Thermal Conductivity of Liquid

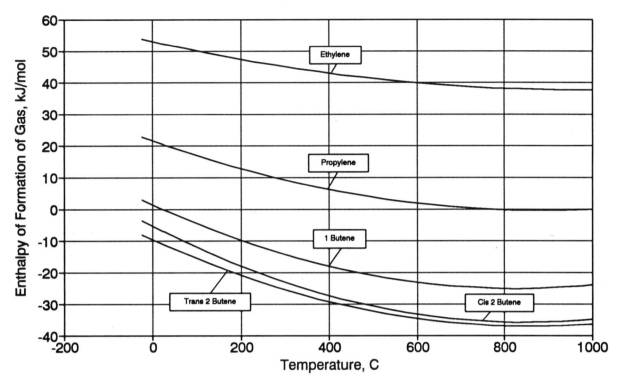

Figure 2-11 Enthalpy of Formation of Gas

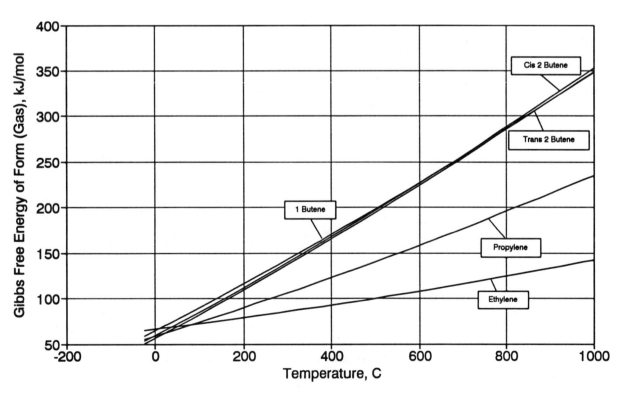

Figure 2-12 Gibbs Free Energy of Formation of Gas

Chapter 3

C₂ TO C₄ ALKYNES

Robert W. Gallant and Carl L. Yaws

PHYSICAL PROPERTIES - Table 3-1

Property data from the literature (1-55,84-93) are given in Table 3-1. Critical constants have been determined experimentally for acetylene and methylacetylene (1-7). Critical pressure and volume are estimated for 1 butyne and 2 butyne (5). Additional property data such as acentric factor, enthalpy of formation, lower explosion limit in air and solubility in water are also available. The DIPPR (Design Institute for Physical Property Research) project (5) and recent data compilations by Yaws and co-workers (44-55) were consulted extensively in preparing the tabulation.

VAPOR PRESSURE - Figure 3-1

Results from the DIPPR project (5) were selected for vapor pressure from very low temperatures to the critical point. Correlation of data for vapor pressure as a function of temperature was accomplished using Equation (1-1). Results from this equation (Antoine-type with extended terms) are in favorable agreement with experimental data. Errors are about 1-5% or less in most cases.

HEAT OF VAPORIZATION - Figure 3-2

The data compilation of Yaws and co-workers (44,52) was selected for heat of vaporization for temperatures ranging from melting point to critical point. The Watson equation, Equation (1-2), was used for correlation of the data as a function of temperature. Reliability of results is good with errors of about 1-5% or less.

LIQUID DENSITY - Figure 3-3

Results from the data compilation of Yaws and co-workers (44,54) were selected for liquid density from low temperatures at the melting point to higher temperatures up to the critical point. A modified Rackett equation, Equation (1-3), was used for correlation of the data as a function of temperature. Results from the correlation are in favorable agreement with data. Deviations are less than 1-2% in most cases.

SURFACE TENSION - Figure 3-4

The data compilation of Yaws and co-workers (44,55) was selected for surface tension for temperatures from melting point to critical point. Using data from the literature, correlation for surface tension as a function of temperature over the full liquid range was achieved by the modified Othmer equation, Equation (1-4). Accuracy is good with errors being about 1-10% or less in most cases.

HEAT CAPACITY - Figures 3-5 and 3-6

Results from the data compilation of Yaws and co-workers (44) were selected for heat capacity of ideal gas. Correlation of data was accomplished using a series expansion in temperature, Equation (1-5). Results are in favorable agreement with data. Errors are about 1% or less in most cases.

Results from Gallant (30) were selected for liquid heat capacity of acetylene. Results from the data compilation of Yaws and co-workers (44) were selected for liquid heat capacity of methylacetylene, 1 butyne and 2 butyne. This selection provided a wider temperature range of coverage than the DIPPR project (5). The coverage applies to temperatures from below the boiling point to temperatures above the boiling point for most of the compounds. Data were correlated with a series expansion in temperature, Equation (1-6). Results are in favorable agreement with the very limited data for 1 butyne. The graphical values should be considered rough estimates.

VISCOSITY - Figures 3-7 and 3-8

The DIPPR project (5) was selected for viscosity of gas. Data for gas viscosity as a function of temperature were correlated using Equation (1-7). Results are in favorable agreement with data. Errors are about 5% or less in most cases.

The DIPPR project (5) was also selected for viscosity of liquid. Temperatures from below the boiling point to temperatures above the boiling point are covered for most of the compounds. Data for liquid viscosity as a function of temperature were correlated using the de Guzman - Andrade equation with extended terms, Equation (1-8). Correlation results and data are in fair agreement with errors being about 5-25% or less.

THERMAL CONDUCTIVITY - Figures 3-9 and 3-10

Results from the DIPPR project (5) were selected for thermal conductivity of gas. Data for gas thermal conductivity as a function of temperature were correlated using the Equation (1-9). Reliability of results is good with errors of about 3-10% or less in most cases.

The estimated results of Gallant (30) were selected for liquid thermal conductivity of acetylene. For the other compounds, results from the DIPPR project (5) were selected. The coverage applies to temperatures from below the boiling point to temperatures above the boiling point for most of the compounds. Data for liquid thermal conductivity as a function of temperature were correlated using a series expansion in temperature, Equation (1-10). Results are in favorable agreement with data. Errors are about 3-10% or less in most cases.

ENTHALPY OF FORMATION - Figure 3-11

The data compilation of Yaws and co-workers (44,45) was for selected enthalpy of formation of ideal gas. Data for enthalpy of formation of the ideal gas is a series expansion in temperature, Equation (1-11). Results from the correlation are in favorable agreement with data.

GIBB'S FREE ENERGY OF FORMATION - Figure 3-12

Results from the data compilation of Yaws and co-workers (44,46) were selected for Gibb's free energy of formation of ideal gas. Data for Gibb's free energy of formation of the ideal gas is a series expansion in temperature, Equation (1-12). Correlation results are in favorable agreement with data.

Table 3-1 Physical Properties

	Acetylene	Methylacetylene	1 Butyne	2 Butyne
1. Name	Acetylene	Methylacetylene	1 Butyne	2 Butyne
2. Formula	C_2H_2	C_3H_4	C_4H_6	C_4H_6
3. Molecular Weight, g/mol	26.038	40.065	54.091	54.091
4. Critical Temperature, K	308.32	402.39	463.65	488.15
5. Critical Pressure, bar	61.391	56.276	47.10	50.800
6. Critical Volume, ml/mol	112.96	164.00	221.00	221.00
7. Critical Compressibility Factor	0.271	0.276	0.270	0.277
8. Acentric Factor	0.1873	0.2161	0.110	0.1305
9. Melting Point, K	192.04	170.45	147.43	240.91
10. Boiling Point @ 1 atm, K	189.00	249.94	281.22	300.13
11. Heat of Vaporization @ BP, kJ/kg	640.37 @ 192.4 K	552.23	461.76	492.73
12. Density of Liquid @ 25 C, g/ml	0.376	0.610	0.640	0.686
13. Surface Tension @ 25 C, dynes/cm	0.977	11.275	17.267	20.016
14. Heat Capacity of Gas @ 25 C, J/g K	1.690	1.514	1.505	1.440
15. Heat Capacity of Liquid @ 25 C, J/g K	3.97	2.77	2.46	2.71
16. Viscosity of Gas @ 25 C, micropoise	102.17	86.35	74.02	75.23
17. Viscosity of Liquid @ 25 C, centipoise	------	0.143	0.203	0.228
18. Thermal Conductivity of Gas @ 25 C, W/m K	0.0213	0.0162	------	0.0142
19. Thermal Conductivity of Liquid @ 25 C, W/m K	0.092	------	------	0.121
20. Enthalpy of Formation of Gas @ 25 C, kJ/mol	226.76	185.45	165.17	146.34
21. Gibbs Free Energy of Formation of Gas @ 25 C, kJ/mol	209.19	194.33	201.91	185.26
22. Flash Point, K	255.37	------	------	------
23. Autoignition Temperature, K	578.15	------	------	------
24. Lower Explosion Limit in Air, vol %	2.5	1.7	------	------
25. Upper Explosion Limit in Air, vol %	80.0	------	------	------
26. Solubility in Water @ 25 C, ppm(wt)	1000	3640	2870	------

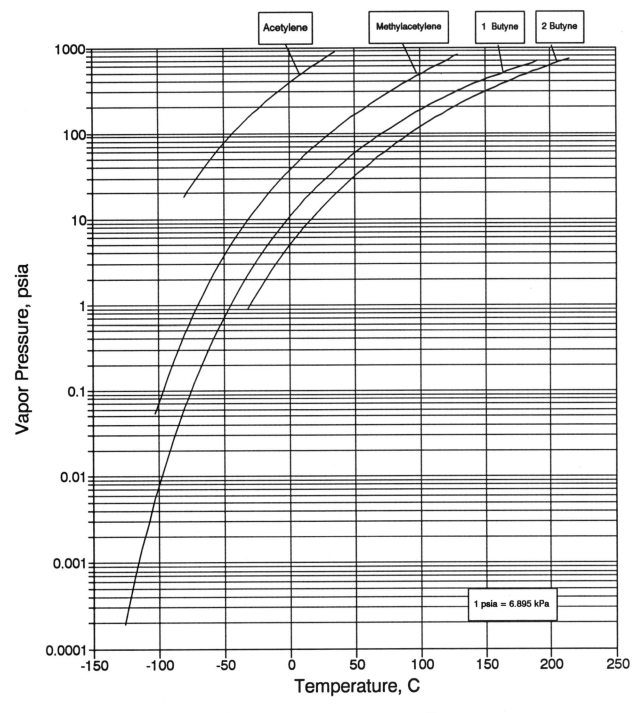

Figure 3-1 Vapor Pressure

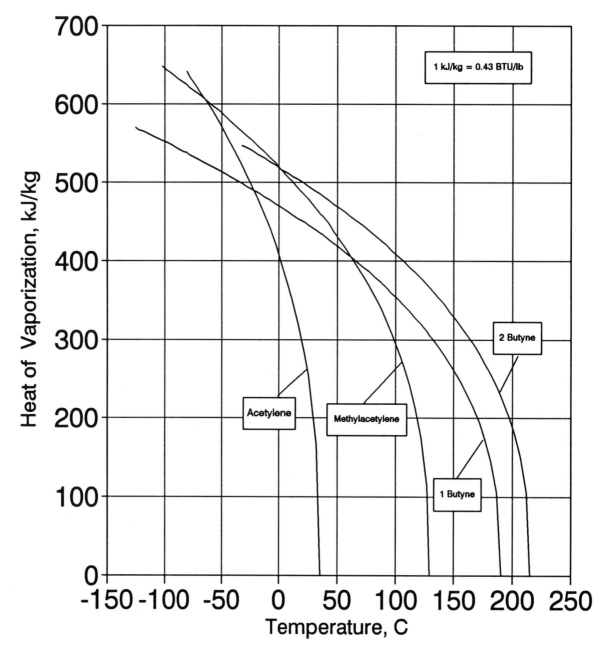

Figure 3-2 Heat of Vaporization

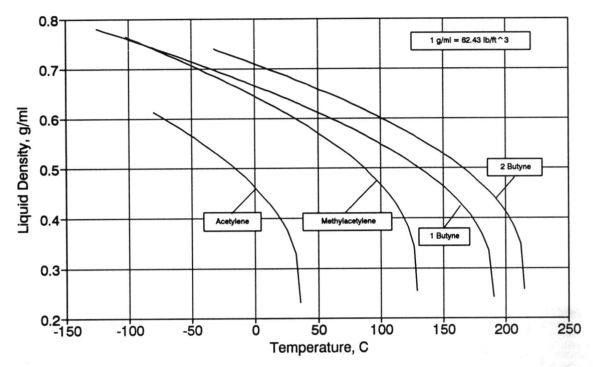

Figure 3-3 Liquid Density

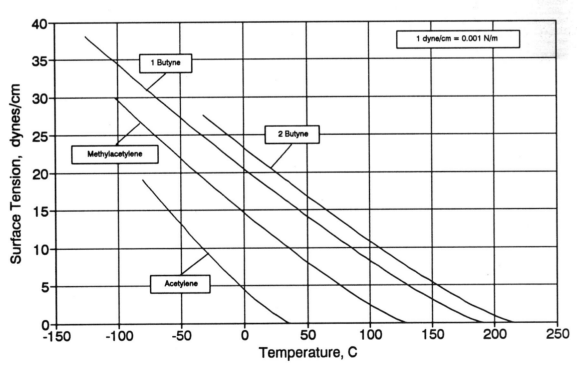

Figure 3-4 Surface Tension

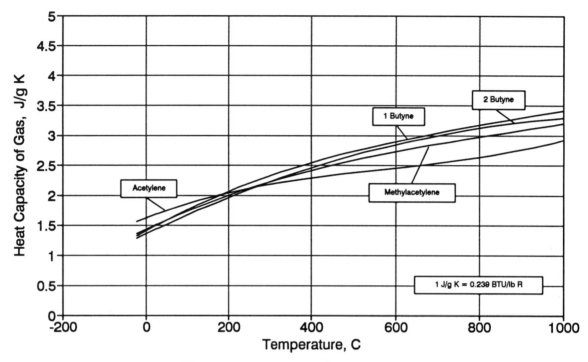

Figure 3-5 Heat Capacity of Gas

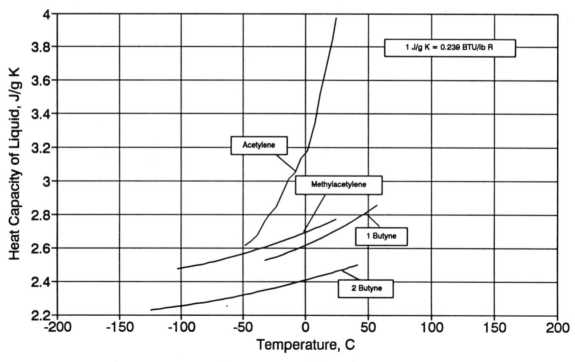

Figure 3-6 Heat Capacity of Liquid

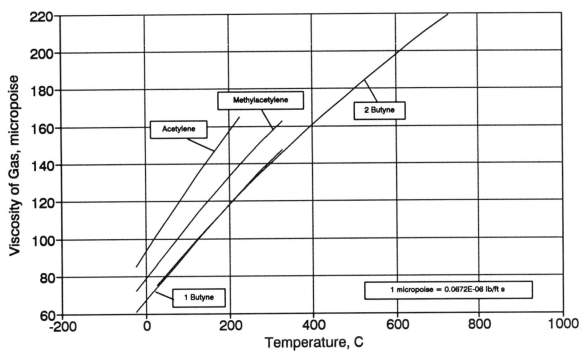

Figure 3-7 Viscosity of Gas

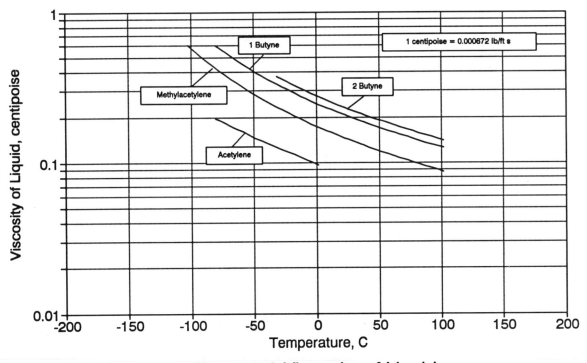

Figure 3-8 Viscosity of Liquid

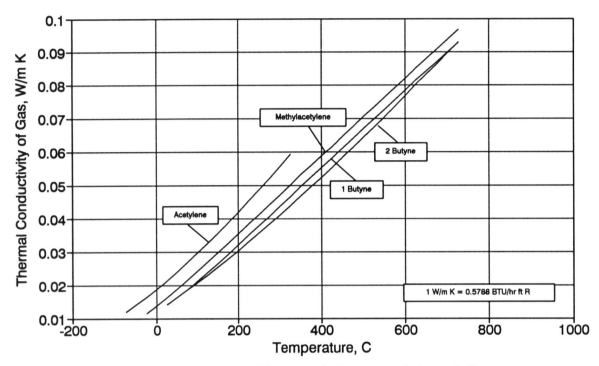

Figure 3-9 Thermal Conductivity of Gas

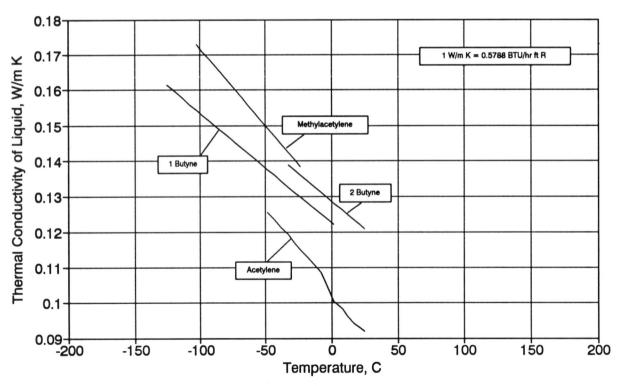

Figure 3-10 Thermal Conductivity of Liquid

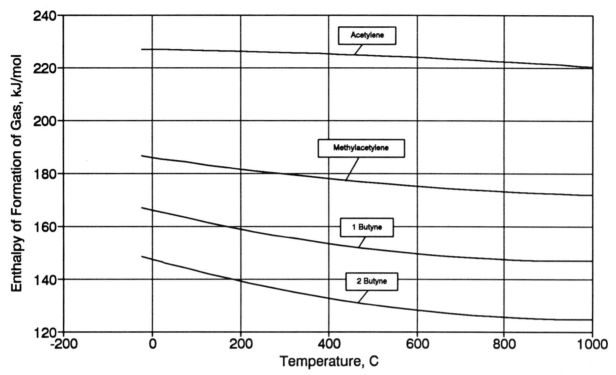

Figure 3-11 Enthalpy of Formation of Gas

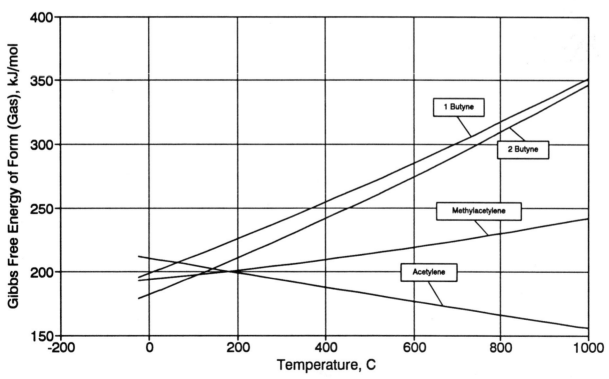

Figure 3-12 Gibbs Free Energy of Formation of Gas

Chapter 4

C₂ TO C₄ DIOLEFINS

Robert W. Gallant and Carl L. Yaws

PHYSICAL PROPERTIES - Table 4-1

Table 4-1 presents property data from the literature (1-55,62,80,88,91-96). Critical temperature for propadiene and all critical constants for 1,3 butadiene have been determined experimentally (1-7). The remaining critical constants are estimated (5). Additional property data such as acentric factor, enthalpy of formation, lower explosion limit in air and solubility in water are also available. The DIPPR (Design Institute for Physical Property Research) project (5) and recent data compilations by Yaws and co-workers (44-55) were consulted extensively in preparing the tabulation.

VAPOR PRESSURE - Figure 4-1

Results from the DIPPR project (5) were selected for vapor pressure from very low temperatures to the critical point. Correlation of data for vapor pressure as a function of temperature was accomplished using Equation (1-1). Results from this equation (Antoine-type with extended terms) are in favorable agreement with experimental data. Errors are about 1-5% or less in most cases.

HEAT OF VAPORIZATION - Figure 4-2

The data compilation of Yaws and co-workers (44,52) was selected for heat of vaporization for temperatures ranging from melting point to critical point. The Watson equation, Equation (1-2), was used for correlation of the data as a function of temperature. Reliability of results is good with errors of about 1-5% or less.

LIQUID DENSITY - Figure 4-3

Results from the data compilation of Yaws and co-workers (44,54) were selected for liquid density from low temperatures at the melting point to higher temperatures up to the critical point. A modified Rackett equation, Equation (1-3), was used for correlation of the data as a function of temperature. Results from the correlation are in favorable agreement with data. Deviations are less than 1-2% in most cases.

SURFACE TENSION - Figure 4-4

The data compilation of Yaws and co-workers (44,55) was selected for surface tension for temperatures from melting point to critical point. Using data from the literature, correlation for surface tension as a function of temperature over the full liquid range was achieved by the modified Othmer equation, Equation (1-4). Accuracy is good with errors being about 1-10% or less in most cases.

HEAT CAPACITY - Figures 4-5 and 4-6

Results from the data compilation of Yaws and co-workers (44) were selected for heat capacity of ideal gas. Correlation of data was accomplished using a series expansion in temperature, Equation (1-5). Results are in favorable agreement with data. Errors are about 1% or less in most cases.

Results from the DIPPR project (5) were selected for heat capacity of liquid. The coverage applies to temperaures from below the boiling point to temperatures above the boiling point for most of the compounds. Data were correlated with a series expansion in temperature, Equation (1-6). Results are in favorable agreement with data. Errors are about 5% or less using the correlation.

VISCOSITY - Figures 4-7 and 4-8

The DIPPR project (5) was selected for viscosity of gas. Data for gas viscosity as a function of temperature were correlated using Equation (1-7). Results are in favorable agreement with data. Errors are about 5% or less in most cases.

The DIPPR project (5) was also selected for viscosity of liquid. Temperaures from below the boiling point to temperatures above the boiling point are covered for most of the compounds. Data for liquid viscosity as a function of temperature were correlated using the de Guzman - Andrade equation with extended terms, Equation (1-8). Correlation results and data are in favorable agreement with errors being about 3-10% or less.

THERMAL CONDUCTIVITY - Figures 4-9 and 4-10

Results from the DIPPR project (5) were selected for thermal conductivity of gas. Data for gas thermal conductivity as a function of temperature were correlated using the Equation (1-9). Reliability of results is good with errors of about 5-10% or less in most cases.

Results from the DIPPR project (5) were selected for thermal conductivity of liquid. The coverage applies to temperaures from below the boiling point to temperatures above the boiling point for most of the compounds. Data for liquid thermal conductivity as a function of temperature were correlated using a series expansion in temperature, Equation (1-10). Results are in favorable agreement with data. Errors are about 5-10% or less in most cases.

ENTHALPY OF FORMATION - Figure 4-11

The data compilation of Yaws and co-workers (44,45) was for selected enthalpy of formation of ideal gas. Data for enthalpy of formation of the ideal gas is a series expansion in temperature, Equation (1-11). Results from the correlation are in favorable agreement with data.

GIBB'S FREE ENERGY OF FORMATION - Figure 4-12

Results from the data compilation of Yaws and co-workers (44,46) were selected for Gibb's free energy of formation of ideal gas. Data for Gibb's free energy of formation of the ideal gas is a series expansion in temperature, Equation (1-12). Correlation results from the correlation are in favorable agreement with data.

Table 4-1 Physical Properties

	Propadiene	1,3 Butadiene	1,2 Butadiene
1. Name	Propadiene	1,3 Butadiene	1,2 Butadiene
2. Formula	C_3H_4	C_4H_6	C_4H_6
3. Molecular Weight, g/mol	40.065	54.091	54.091
4. Critical Temperature, K	393.15	425.37	444.00
5. Critical Pressure, bar	54.70	43.299	45.00
6. Critical Volume, ml/mol	162.00	220.84	219.00
7. Critical Compressibility Factor	0.271	0.270	0.267
8. Acentric Factor	0.1594	0.1932	0.2509
9. Melting Point, K	136.87	164.25	136.95
10. Boiling Point @ 1 atm, K	238.65	268.74	284.00
11. Heat of Vaporization @ BP, kJ/kg	467.67	415.33	448.63
12. Density of Liquid @ 25 C, g/ml	0.582	0.615	0.646
13. Surface Tension @ 25 C, dynes/cm	8.73	12.50	15.97
14. Heat Capacity of Gas @ 25 C, J/g K	1.473	1.471	1.481
15. Heat Capacity of Liquid @ 25 C, J/g K	------	2.285	2.299
16. Viscosity of Gas @ 25 C, micropoise	83.02	86.62	76.28
17. Viscosity of Liquid @ 25 C, centipoise	0.142	0.146	0.190
18. Thermal Conductivity of Gas @ 25 C, W/m K	0.0161	0.0157	0.0155
19. Thermal Conductivity of Liquid @ 25 C, W/m K	------	------	------
20. Enthalpy of Formation of Gas @ 25 C, kJ/mol	192.13	110.08	162.19
21. Gibbs Free Energy of Formation of Gas @ 25 C, kJ/mol	202.27	150.50	198.28
22. Flash Point, K	------	------	------
23. Autoignition Temperature, K	------	702.04	------
24. Lower Explosion Limit in Air, vol %	2.1	2.0	2.0
25. Upper Explosion Limit in Air, vol %	------	11.5	12.0
26. Solubility in Water @ 25 C, ppm(wt)	------	735	------

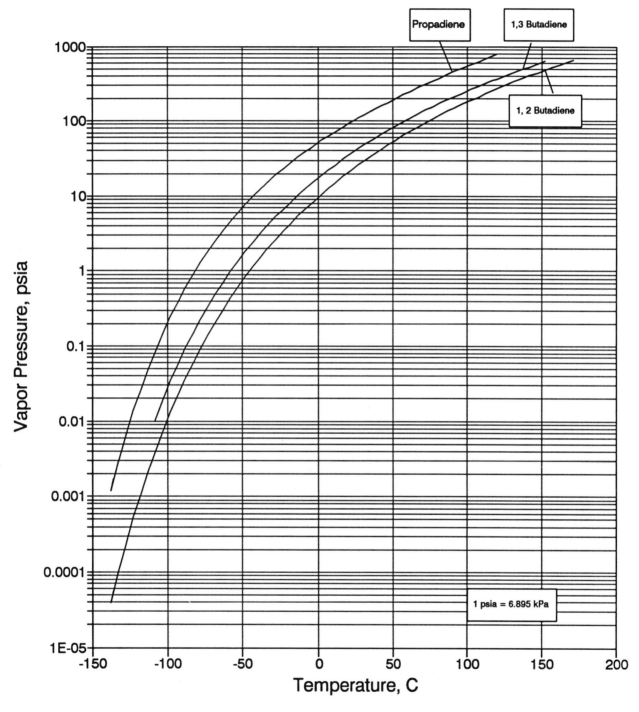

Figure 4-1 Vapor Pressure

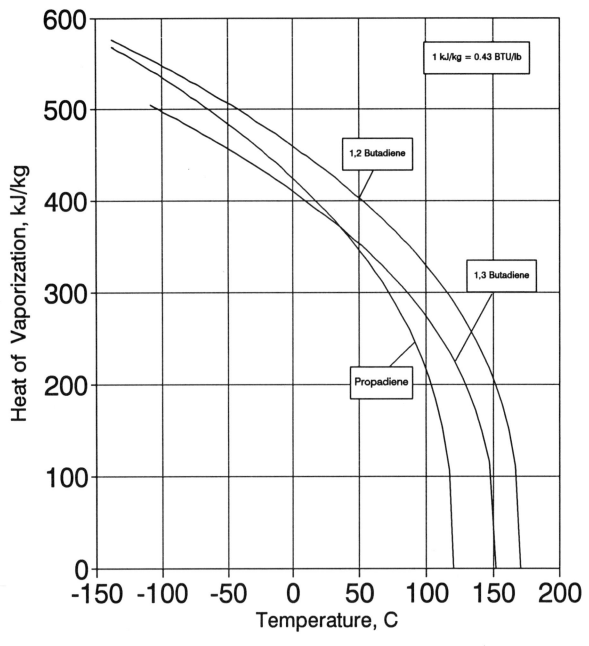

Figure 4-2 Heat of Vaporization

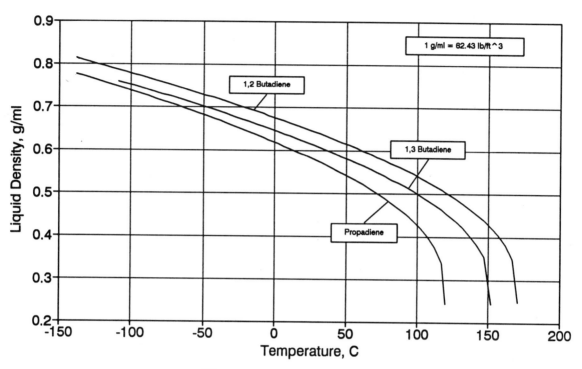

Figure 4-3 Liquid Density

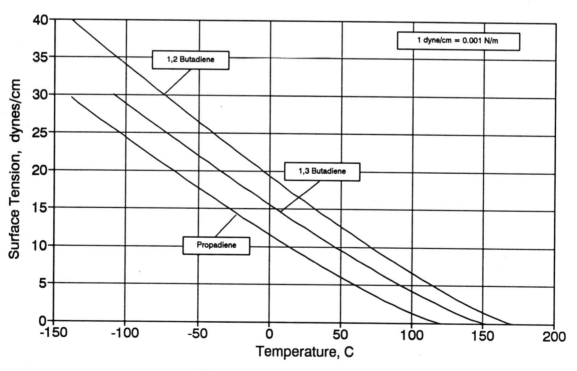

Figure 4-4 Surface Tension

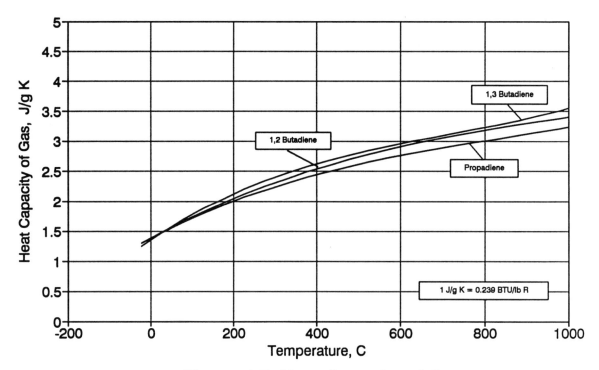

Figure 4-5 Heat Capacity of Gas

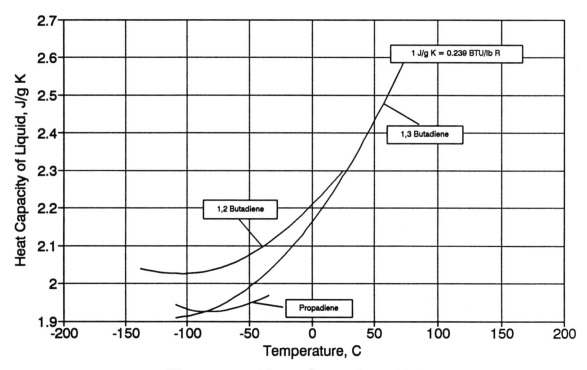

Figure 4-6 Heat Capacity of Liquid

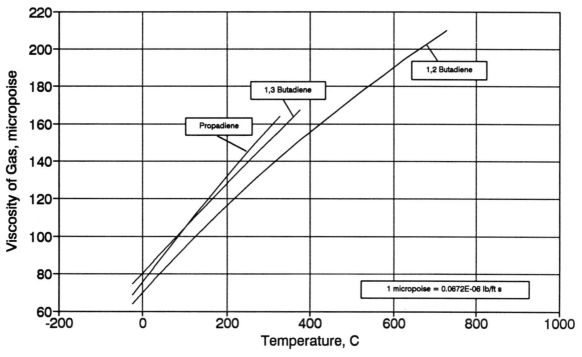

Figure 4-7 Viscosity of Gas

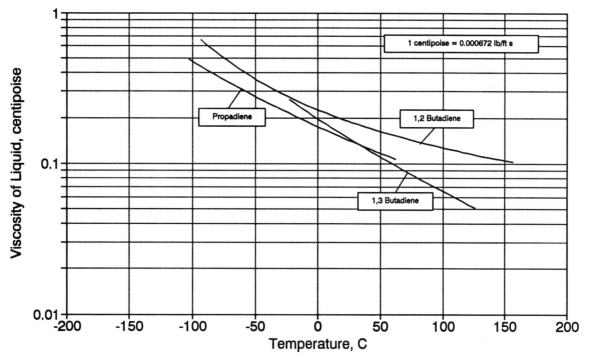

Figure 4-8 Viscosity of Liquid

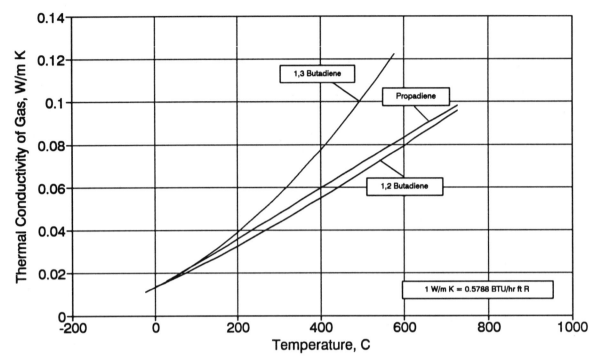

Figure 4-9 Thermal Conductivity of Gas

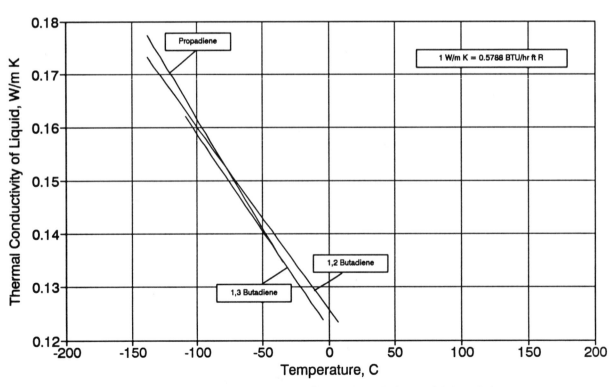

Figure 4-10 Thermal Conductivity of Liquid

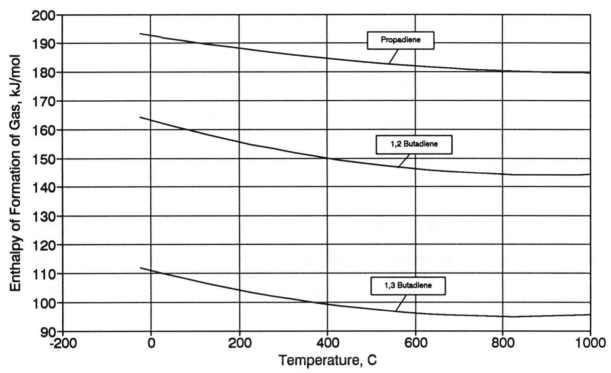

Figure 4-11 Enthalpy of Formation of Gas

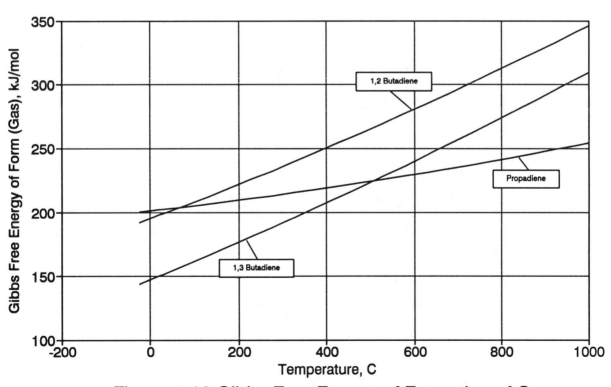

Figure 4-12 Gibbs Free Energy of Formation of Gas

Chapter 5

CHLORINATED METHANES

Robert W. Gallant, Carl L. Yaws and Xiang Pan

PHYSICAL PROPERTIES - Table 5-1

Property data from the literature (1-55,97-108) are given in Table 5-1. Critical volume for dichloromethane is estimated (5). The remaining critical constants have been determined experimentally (1-7). Additional property data such as acentric factor, enthalpy of formation, lower explosion limit in air and solubility in water are also available. The DIPPR (Design Institute for Physical Property Research) project (5) and recent data compilations by Yaws and co-workers (44-55) were consulted extensively in preparing the tabulation.

VAPOR PRESSURE - Figure 5-1

Results from the DIPPR project (5) were selected for vapor pressure from very low temperatures to the critical point. Correlation of data for vapor pressure as a function of temperature was accomplished using Equation (1-1). Results from this equation (Antoine-type with extended terms) are in favorable agreement with experimental data. Errors are about 1-5% or less in most cases.

HEAT OF VAPORIZATION - Figure 5-2

The data compilation of Yaws and co-workers (44,52) was selected for heat of vaporization for temperatures ranging from melting point to critical point. The Watson equation, Equation (1-2), was used for correlation of the data as a function of temperature. Reliability of results is good with errors of about 1-5% or less.

LIQUID DENSITY - Figure 5-3

Results from the data compilation of Yaws and co-workers (44,54) were selected for liquid density from low temperatures at the melting point to higher temperatures up to the critical point. A modified Rackett equation, Equation (1-3), was used for correlation of the data as a function of temperature. Results from the correlation are in favorable agreement with data. Deviations are less than 1-2% in most cases.

SURFACE TENSION - Figure 5-4

The data compilation of Yaws and co-workers (44,55) was selected for surface tension for temperatures from melting point to critical point. Using data from the literature, correlation for surface tension as a function of temperature over the full liquid range was achieved by the modified Othmer equation, Equation (1-4). Accuracy is good with errors being about 1-10% or less in most cases.

HEAT CAPACITY - Figures 5-5 and 5-6

Results from the data compilation of Yaws and co-workers (44) were selected for heat capacity of ideal gas. Correlation of data was accomplished using a series expansion in temperature, Equation (1-5). Results are in favorable agreement with data. Errors are about 1% or less in most cases.

Results from the DIPPR project (5) were selected for heat capacity of liquid. The coverage applies to temperatures from below the boiling point to temperatures above the boiling point for most of the compounds. Data were correlated with a series expansion in temperature, Equation (1-6). Results are in favorable agreement with data. Errors are about 5% or less using the correlation.

VISCOSITY - Figures 5-7 and 5-8

The DIPPR project (5) was selected for viscosity of gas. Data for gas viscosity as a function of temperature were correlated using Equation (1-7). Results are in favorable agreement with data. Errors are about 3-10% or less in most cases.

The DIPPR project (5) was also selected for viscosity of liquid. Temperatures from below the boiling point to temperatures above the boiling point are covered for most of the compounds. Data for liquid viscosity as a function of temperature were correlated using the de Guzman - Andrade equation with extended terms, Equation (1-8). Correlation results and data are in favorable agreement with errors being about 1-5% or less.

THERMAL CONDUCTIVITY - Figures 5-9 and 5-10

Results from the DIPPR project (5) were selected for thermal conductivity of gas. Data for gas thermal conductivity as a function of temperature were correlated using the Equation (1-9). Reliability of results is good with errors of about 5-10% or less in most cases.

Results from the DIPPR project (5) were selected for thermal conductivity of liquid. The coverage applies to temperatures from below the boiling point to temperatures above the boiling point for most of the compounds. Data for liquid thermal conductivity as a function of temperature were correlated using a series expansion in temperature, Equation (1-10). Results are in favorable agreement with data. Errors are about 5% or less in most cases.

ENTHALPY OF FORMATION - Figure 5-11

The data compilation of Yaws and co-workers (44,45) was selected for enthalpy of formation of ideal gas. Data for enthalpy of formation of the ideal gas is a series expansion in temperature, Equation (1-11). Results from the correlation are in favorable agreement with data.

GIBB'S FREE ENERGY OF FORMATION - Figure 5-12

Results from the data compilation of Yaws and co-workers (44,46) were selected for Gibb's free energy of formation of ideal gas. Data for Gibb's free energy of formation of the ideal gas is a series expansion in temperature, Equation (1-12). Correlation results are in favorable agreement with data.

Table 5-1 Physical Properties

1. Name	Metyl Chloride	Methylene Chloride	Chloroform	Carbon Tetrachloride
2. Formula	CH3Cl	CH2Cl2	CHCl3	CCl4
3. Molecular Weight, g/mol	50.488	84.933	119.378	153.823
4. Critical Temperature, K	416.25	510.00	536.40	556.35
5. Critical Pressure, bar	66.793	60.795	54.716	45.596
6. Critical Volume, ml/mol	139.00	185.00	239.00	276.00
7. Critical Compressibility Factor	0.268	0.265	0.293	0.272
8. Acentric Factor	0.1529	0.1916	0.2129	0.1926
9. Melting Point, K	175.42	178.01	209.63	250.33
10. Boiling Point @ 1 atm, K	248.93	312.90	334.33	349.79
11. Heat of Vaporization @ BP, kJ/kg	424.60	329.76	248.83	195.03
12. Density of Liquid @ 25 C, g/ml	0.905	1.319	1.481	1.584
13. Surface Tension @ 25 C, dynes/cm	15.41	27.07	26.68	26.44
14. Heat Capacity of Gas @ 25 C, J/g K	0.81	0.60	0.55	0.55
15. Heat Capacity of Liquid @ 25 C, J/g K	1.61	1.19	0.96	0.85
16. Viscosity of Gas @ 25 C, micropoise	109.52	103.03	102.25	99.87
17. Viscosity of Liquid @ 25 C, centipoise	0.1725	0.4134	0.5387	0.9048
18. Thermal Conductivity of Gas @ 25 C, W/m K	0.0107	0.0074	0.0066	0.0065
19. Thermal Conductivity of Liquid @ 25 C, W/m K	0.1579	0.1392	0.1172	0.1023
20. Enthalpy of Formation of Gas @ 25 C, kJ/mol	-86.34	-95.42	-101.30	-100.45
21. Gibbs Free Energy of Formation of Gas @ 25 C, kJ/mol	-63.01	-68.95	-68.57	-56.66
22. Flash Point, K	------	269.00	------	------
23. Autoignition Temperature, K	905.37	935.37	------	------
24. Lower Explosion Limit in Air, vol %	10.7	15.5	------	------
25. Upper Explosion Limit in Air, vol %	12.4	66.0	------	------
26. Solubility in Water @ 25 C, ppm(wt)	5900	1938	7500	785.7

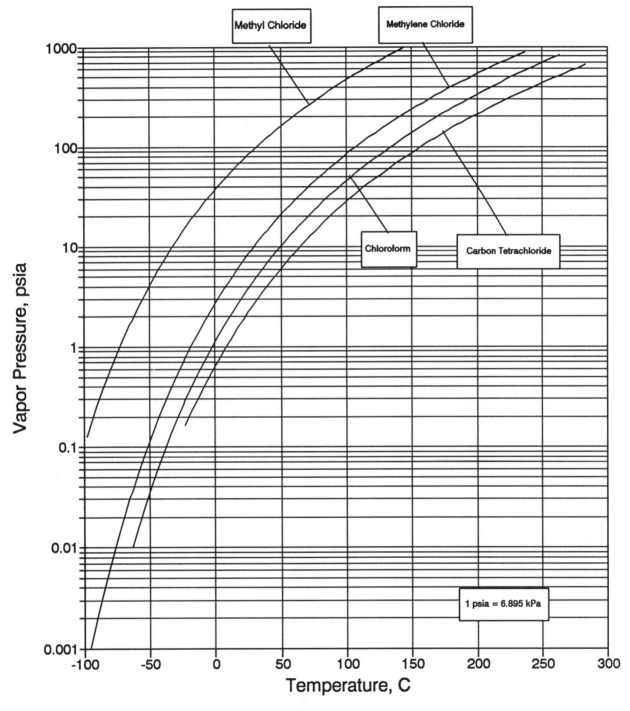

Figure 5-1 Vapor Pressure

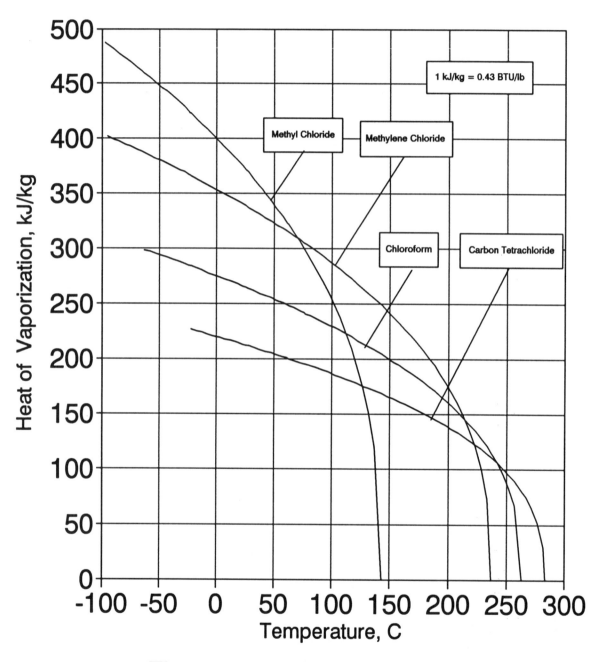

Figure 5-2 Heat of Vaporization

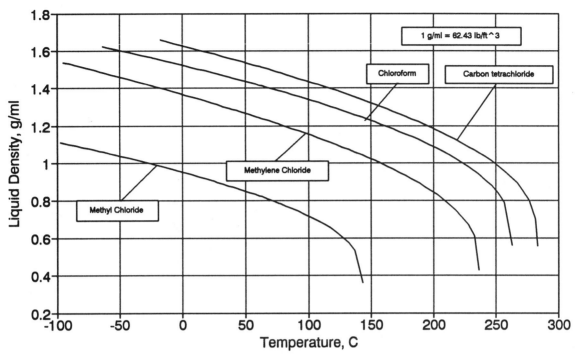

Figure 5-3 Liquid Density

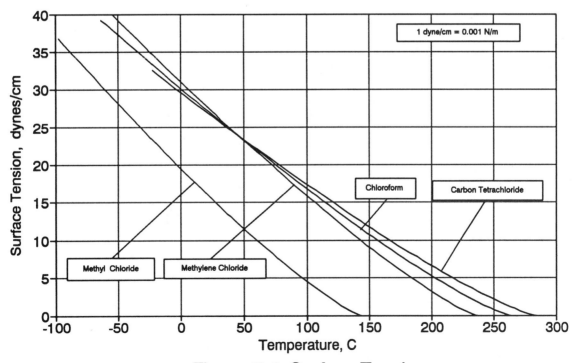

Figure 5-4 Surface Tension

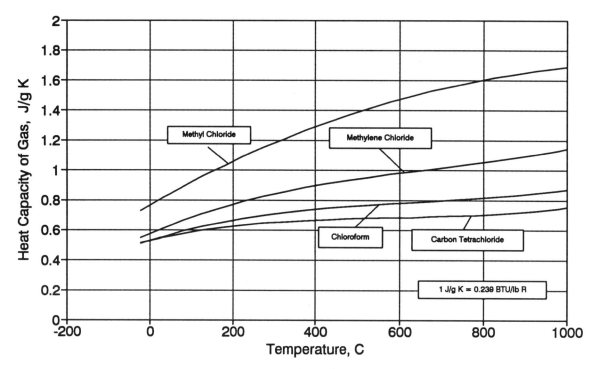

Figure 5-5 Heat Capacity of Gas

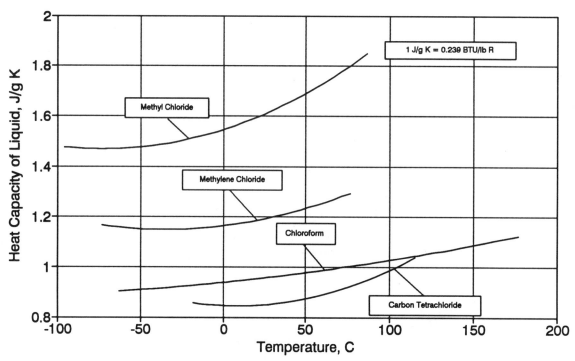

Figure 5-6 Heat Capacity of Liquid

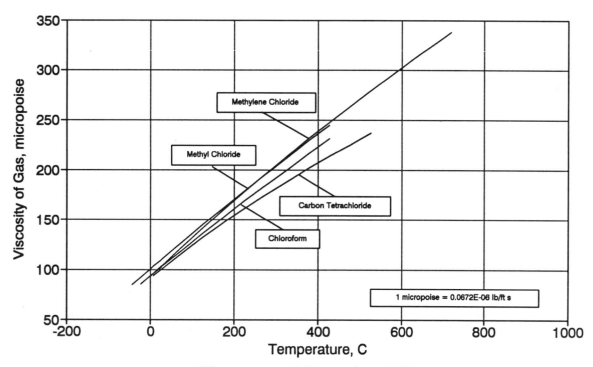

Figure 5-7 Viscosity of Gas

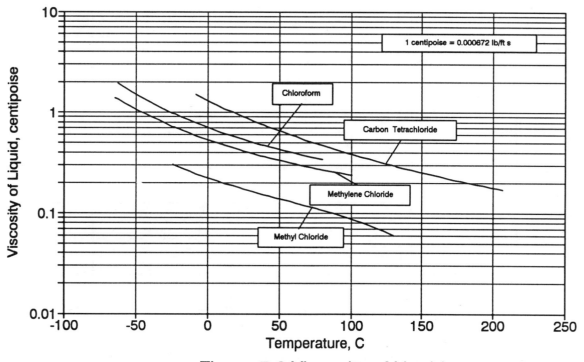

Figure 5-8 Viscosity of Liquid

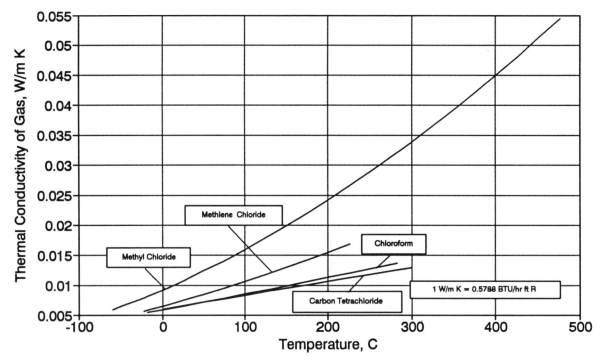

Figure 5-9 Thermal Conductivity of Gas

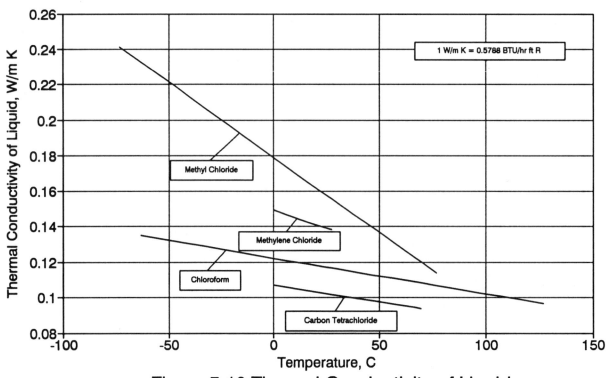

Figure 5-10 Thermal Conductivity of Liquid

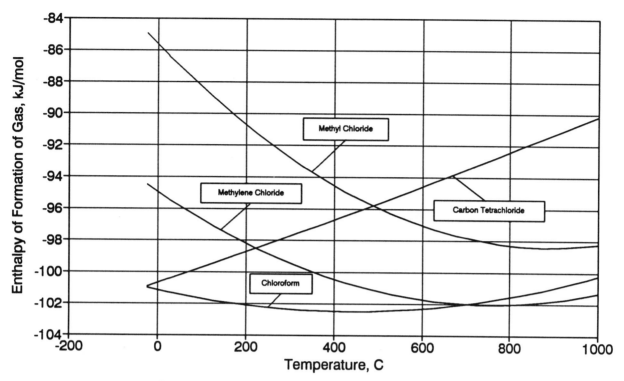

Figure 5-11 Enthalpy of Formation of Gas

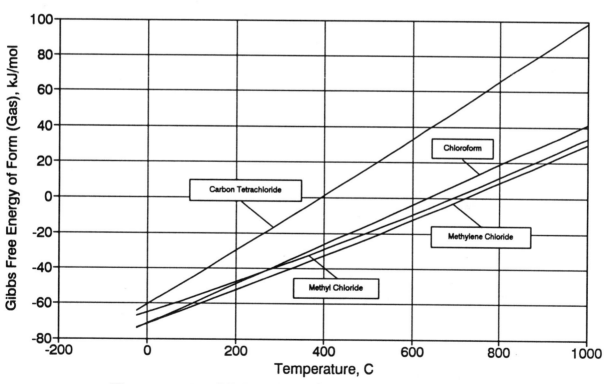

Figure 5-12 Gibbs Free Energy of Formation of Gas

Chapter 6

CHLORINATED ETHYLENES

Robert W. Gallant and Carl L. Yaws

PHYSICAL PROPERTIES - Table 6-1

Property data from the literature (1-55,100,109-118) are given in Table 6-1. Since results for vinylidene chloride are not available in the DIPPR project, critical constants for vinylidene chloride were selected from Yaws (44,47). Critical constants for the remaining compounds were selected from the DIPPR project (5). Additional property data such as acentric factor, enthalpy of formation, lower explosion limit in air and solubility in water are also available. The DIPPR (Design Institute for Physical Property Research) project (5) and recent data compilations by Yaws and co-workers (44-55) were consulted extensively in preparing the tabulation.

VAPOR PRESSURE - Figure 6-1

For vinylidene chloride, results were selected from Yaws and co-workers (44,48). For the remaining compounds, results from the DIPPR project (5) were selected for vapor pressure from very low temperatures to the critical point. Correlation of data for vapor pressure as a function of temperature was accomplished using Equation (1-1). Results from this equation (Antoine-type with extended terms) are in favorable agreement with experimental data. Errors are about 1-5% or less in most cases.

HEAT OF VAPORIZATION - Figure 6-2

The data compilation of Yaws and co-workers (44,52) was selected for heat of vaporization for temperatures ranging from melting point to critical point. The Watson equation, Equation (1-2), was used for correlation of the data as a function of temperature. Reliability of results is good with errors of about 5% or less in most cases.

LIQUID DENSITY - Figure 6-3

Results from the data compilation of Yaws and co-workers (44,54) were selected for liquid density from low temperatures at the melting point to higher temperatures up to the critical point. A modified Rackett equation, Equation (1-3), was used for correlation of the data as a function of temperature. Results from the correlation are in favorable agreement with data. Deviations are less than 2% in most cases.

SURFACE TENSION - Figure 6-4

The data compilation of Yaws and co-workers (44,55) was selected for surface tension for temperatures from melting point to critical point. Using data from the literature, correlation for surface tension as a function of temperature over the full liquid range was achieved by the modified Othmer equation, Equation (1-4). Accuracy is good with errors being about 1-10% or less in most cases.

HEAT CAPACITY - Figures 6-5 and 6-6

Results from the data compilation of Yaws and co-workers (44) were selected for heat capacity of ideal gas. Correlation of data was accomplished using a series expansion in temperature, Equation (1-5). Results are in favorable agreement with data. Errors are about 1% or less in most cases.

For vinylidene chloride, results were selected from Yaws and co-workers (44). For the remaining compounds, results from the DIPPR project (5) were selected for liquid heat capacity. The coverage applies to temperatures from below the boiling point to temperatures above the boiling point for most of the compounds. Data were correlated with a series expansion in temperature, Equation (1-6). Correlation results are in favorable agreement with data.

VISCOSITY - Figures 6-7 and 6-8

Results from the DIPPR project (5) were selected for viscosity of gas. Since the chemical structure of vinylidene chloride is between that of vinyl chloride and trichloroethylene, the values for vinylidene chloride were estimated from values for vinyl chloride and trichloroethylene. In the absence of data, this estimate should be considered a rough approximation. Data for gas viscosity as a function of temperature were correlated using Equation (1-7). Results are in favorable agreement with data. Errors are about 10% or less in most cases.

For vinylidene chloride, results were selected from the original work of Gallant (30). For the remaining compounds, results from the DIPPR project (5) were selected for viscosity of liquid. Temperatures from below the boiling point to temperatures above the boiling point are covered for most of the compounds. Data for liquid viscosity as a function of temperature were correlated using the de Guzman - Andrade equation with extended terms, Equation (1-8). Correlation results and data are in favorable agreement with errors being about 3-5% or less.

THERMAL CONDUCTIVITY - Figures 6-9 and 6-10

Results from the DIPPR project (5) were selected for thermal conductivity of gas. Data for gas thermal conductivity as a function of temperature were correlated using the Equation (1-9). Reliability of results is good with errors of about 5-10% or less in most cases.

Results from the DIPPR project (5) were selected for liquid thermal conductivity. The coverage applies to temperatures from below the boiling point to temperatures above the boiling point for most of the compounds. Data for liquid thermal conductivity as a function of temperature were correlated using a series expansion in temperature, Equation (1-10). Results are in fair agreement with data. Errors are about 3-25% or less in most cases.

Since the chemical structure of vinylidene chloride is between that of vinyl chloride and trichloroethylene, the values for vinylidene chloride were estimated from values for vinyl chloride and trichloroethylene. In the absence of data, this estimate for gas and liquid thermal conductivity of vinylidene chloride should be considered a rough approximation.

ENTHALPY OF FORMATION - Figure 6-11

The data compilation of Yaws and co-workers (44,45) was for selected enthalpy of formation of ideal gas. Data for enthalpy of formation of the ideal gas is a series expansion in temperature, Equation (1-11). Results from the correlation are in favorable agreement with data.

GIBB'S FREE ENERGY OF FORMATION - Figure 6-12

Results from the data compilation of Yaws and co-workers (44,46) were selected for Gibb's free energy of formation of ideal gas. Data for Gibb's free energy of formation of the ideal gas is a series expansion in temperature, Equation (1-12). Correlation results are in favorable agreement with data.

Table 6-1 Physical Properties

	Vinyl Chloride	Vinylidene Chloride	Trichloro-ethylene	Perchloro-ethylene
1. Name	Vinyl Chloride	Vinylidene Chloride	Trichloro-ethylene	Perchloro-ethylene
2. Formula	C_2H_3Cl	$C_2H_2Cl_2$	C_2HCl_3	C_2Cl_4
3. Molecular Weight, g/mol	62.499	96.944	131.389	165.834
4. Critical Temperature, K	432.00	489.00	571.00	620.00
5. Critical Pressure, bar	56.70	46.80	49.10	44.90
6. Critical Volume, ml/mol	179.00	219.00	256.00	248.00
7. Critical Compressibility Factor	0.283	0.252	0.265	0.216
8. Acentric Factor	0.1048	0.179	0.2159	0.2210
9. Melting Point, K	119.36	155.90	188.40	250.80
10. Boiling Point @ 1 atm, K	259.78	304.70	360.10	394.40
11. Heat of Vaporization @ BP, kJ/kg	355.92	311.26	239.31	209.34
12. Density of Liquid @ 25 C, g/ml	0.907	1.269	1.453	1.611
13. Surface Tension @ 25 C, dynes/cm	16.22	22.28	28.37	31.70
14. Heat Capacity of Gas @ 25 C, J/g K	0.858	0.691	0.611	0.573
15. Heat Capacity of Liquid @ 25 C, J/g K	1.375	1.280	0.947	0.864
16. Viscosity of Gas @ 25 C, micropoise	102.41	99.26	96.12	85.34
17. Viscosity of Liquid @ 25 C, centipoise	0.170	0.450	0.545	0.845
18. Thermal Conductivity of Gas @ 25 C, W/m K	0.0121	0.0092	0.0062	0.0054
19. Thermal Conductivity of Liquid @ 25 C, W/m K	------	0.107	0.116	0.110
20. Enthalpy of Formation of Gas @ 25 C, kJ/mol	28.41	2.34	-9.65	-12.14
21. Gibbs Free Energy of Formation of Gas @ 25 C, kJ/mol	43.03	25.33	16.06	22.67
22. Flash Point, K	195.37	------	305.37	------
23. Autoignition Temperature, K	745.37	730.15	683.15	------
24. Lower Explosion Limit in Air, vol %	3.6	5.6	12.5	------
25. Upper Explosion Limit in Air, vol %	33.0	11.4	90.0	------
26. Solubility in Water @ 25 C, ppm(wt)	2697	3345	1100	150

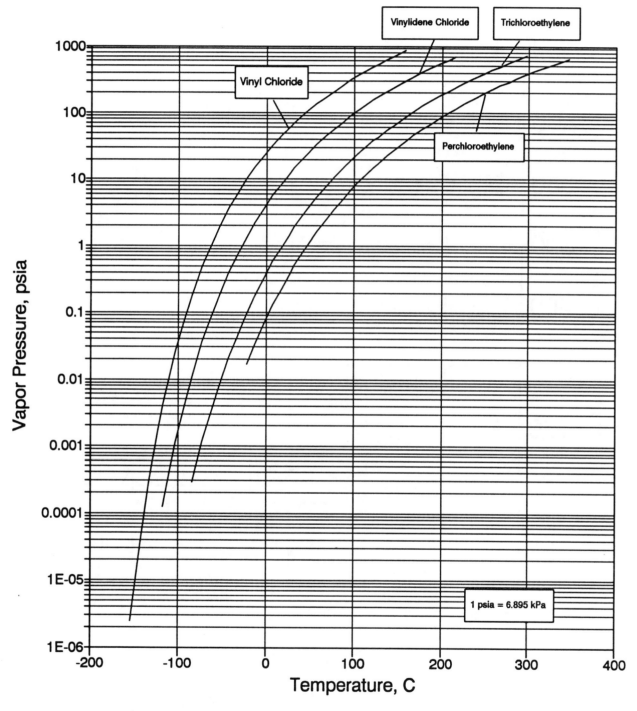

Figure 6-1 Vapor Pressure

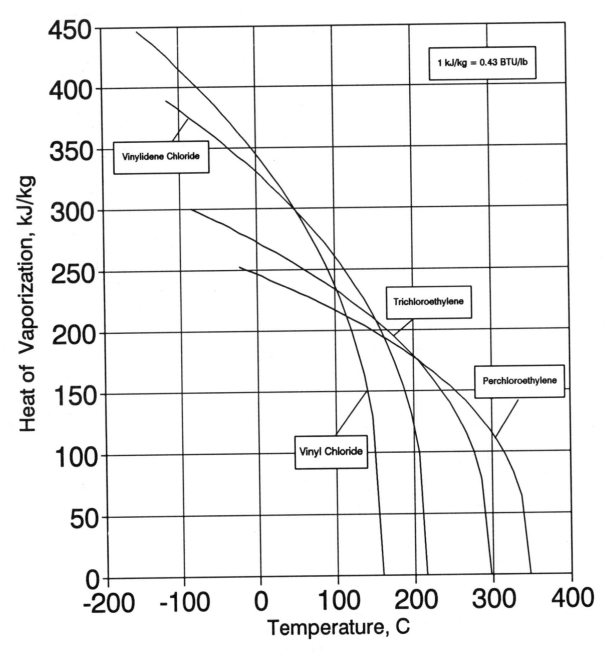

Figure 6-2 Heat of Vaporization

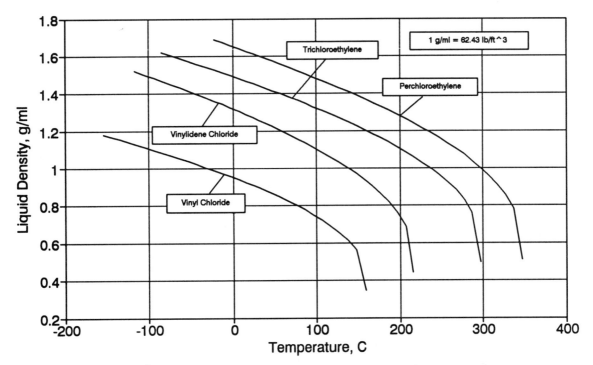

Figure 6-3 Liquid Density

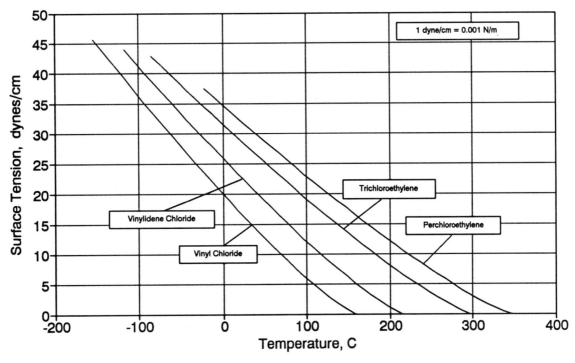

Figure 6-4 Surface Tension

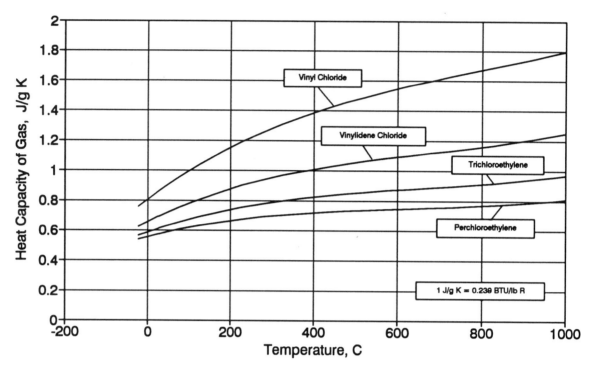

Figure 6-5 Heat Capacity of Gas

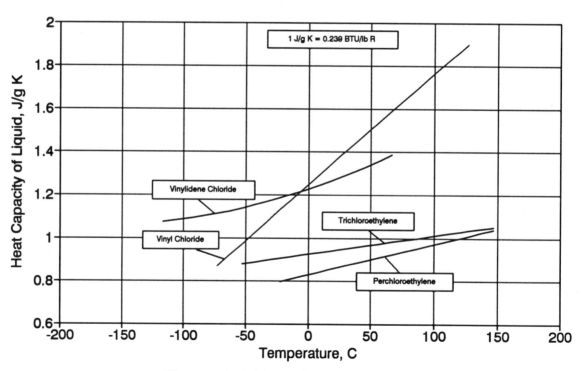

Figure 6-6 Heat Capacity of Liquid

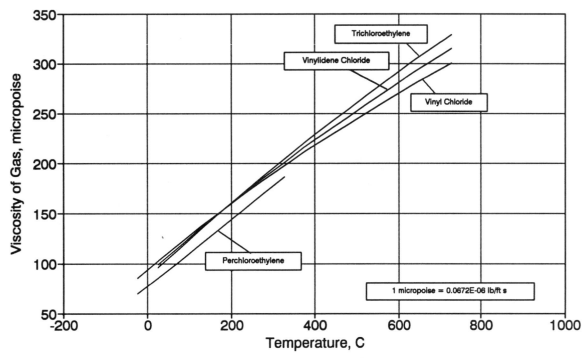

Figure 6-7 Viscosity of Gas

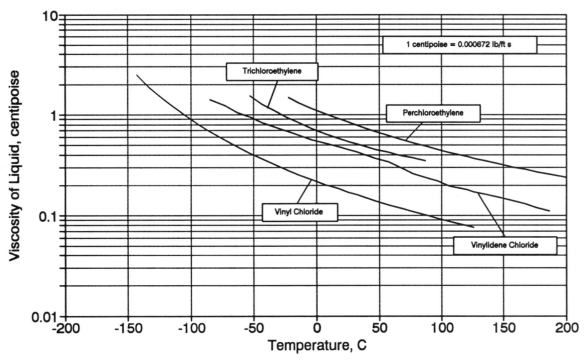

Figure 6-8 Viscosity of Liquid

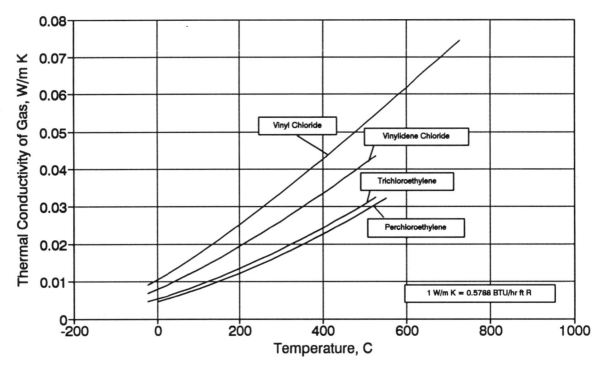

Figure 6-9 Thermal Conductivity of Gas

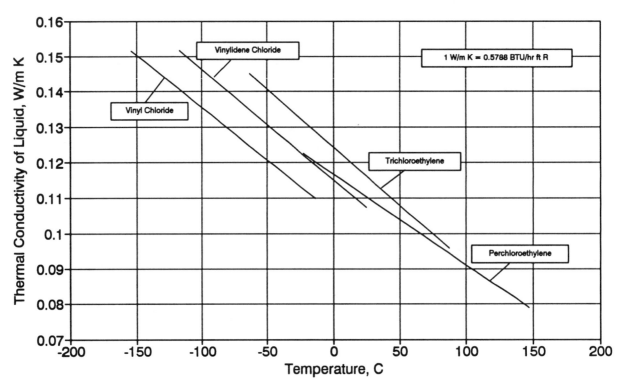

Figure 6-10 Thermal Conductivity of Liquid

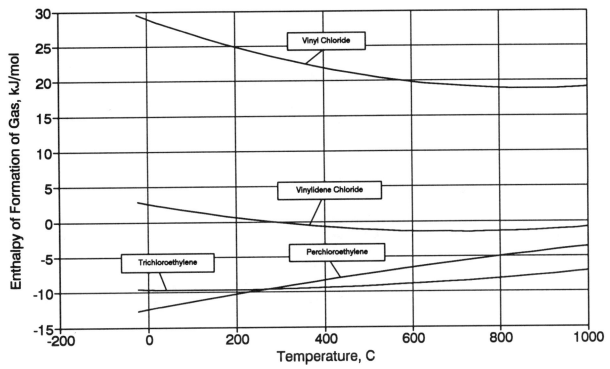

Figure 6-11 Enthalpy of Formation of Gas

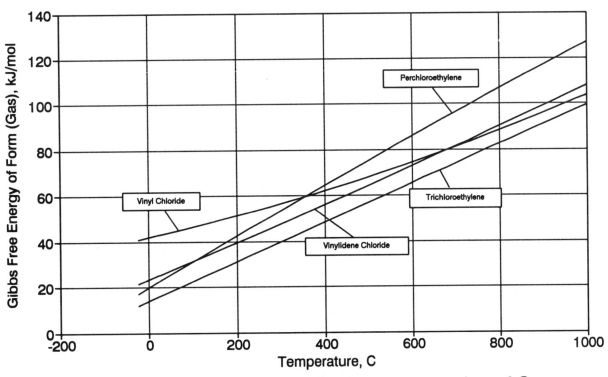

Figure 6-12 Gibbs Free Energy of Formation of Gas

Chapter 7

CHLORINATED ALIPHATICS

Robert W. Gallant, Carl L. Yaws and Xiang Pan

PHYSICAL PROPERTIES - Table 7-1

Property data from the literature (1-55,100,105,111-115,119-125) are given in Table 7-1. Since results for propyl chloride are not available in the DIPPR project, critical constants for propyl chloride were selected from Yaws (44,47). Critical constants for the remaining compounds were selected from the DIPPR project (5). Additional property data such as acentric factor, enthalpy of formation, lower explosion limit in air and solubility in water are also available. The DIPPR (Design Institute for Physical Property Research) project (5) and recent data compilations by Yaws and co-workers (44-55) were consulted extensively in preparing the tabulation.

VAPOR PRESSURE - Figure 7-1

For propyl chloride, results were selected from Yaws and co-workers (44,48). For the remaining compounds, results from the DIPPR project (5) were selected for vapor pressure from very low temperatures to the critical point. Correlation of data for vapor pressure as a function of temperature was accomplished using Equation (1-1). Results from this equation (Antoine-type with extended terms) are in favorable agreement with experimental data. Errors are about 1-5% or less in most cases.

HEAT OF VAPORIZATION - Figure 7-2

The data compilation of Yaws and co-workers (44,52) was selected for heat of vaporization for temperatures ranging from melting point to critical point. The Watson equation, Equation (1-2), was used for correlation of the data as a function of temperature. Reliability of results is good with errors of about 5% or less in most cases.

LIQUID DENSITY - Figure 7-3

Results from the data compilation of Yaws and co-workers (44,54) were selected for liquid density from low temperatures at the melting point to higher temperatures up to the critical point. A modified Rackett equation, Equation (1-3), was used for correlation of the data as a function of temperature. Results from the correlation are in favorable agreement with data. Deviations are less than 2% in most cases.

SURFACE TENSION - Figure 7-4

The data compilation of Yaws and co-workers (44,55) was selected for surface tension for temperatures from melting point to critical point. Using data from the literature, correlation for surface tension as a function of temperature over the full liquid range was achieved by the modified Othmer equation, Equation (1-4). Accuracy is good with errors being about 1-10% or less in most cases.

HEAT CAPACITY - Figures 7-5 and 7-6

Results from the data compilation of Yaws and co-workers (44) were selected for heat capacity of ideal gas. Correlation of data was accomplished using a series expansion in temperature, Equation (1-5). Results are in favorable agreement with data. Errors are about 1% or less in most cases.

Results from Yaws and co-workers (44) were selected for liquid heat capacity. The coverage applies to temperatures from below the boiling point to temperatures above the boiling point for most of the compounds. Data were correlated with a series expansion in temperature, Equation (1-6). Correlation results are in favorable agreement with data.

VISCOSITY - Figures 7-7 and 7-8

For propyl chloride, results were selected from the original work of Gallant (30). For the remaining compounds results from the DIPPR project (5) were selected for viscosity of gas. Data for gas viscosity as a function of temperature were correlated using Equation (1-7). Results are in favorable agreement with data. Errors are about 3-10% or less in most cases.

Results for propyl chloride were selected from the original work of Gallant (30). Results for the remaining compounds were selected from the DIPPR project (5) for viscosity of liquid. Temperatures from below the boiling point to temperatures above the boiling point are covered for most of the compounds. Data for liquid viscosity as a function of temperature were correlated using the de Guzman - Andrade equation with extended terms, Equation (1-8). Correlation results and data are in favorable agreement with errors being about 3-10% or less.

THERMAL CONDUCTIVITY - Figures 7-9 and 7-10

For propyl chloride, results were selected from the original work of Gallant (30). For the remaining compounds, results from the DIPPR project (5) were selected for thermal conductivity of gas. Data for gas thermal conductivity as a function of temperature were correlated using the Equation (1-9). Reliability of results is good with errors of about 10% or less in most cases.

Results for propyl chloride were selected from the original work of Gallant (30). Results for the remaining compounds were selected from the DIPPR project (5) for liquid thermal conductivity. The coverage applies to temperatures from below the boiling point to temperatures above the boiling point for most of the compounds. Data for liquid thermal conductivity as a function of temperature were correlated using a series expansion in temperature, Equation (1-10). Results are in fair agreement with data. Errors are about 3-5% or less in most cases.

ENTHALPY OF FORMATION - Figure 7-11

The data compilation of Yaws and co-workers (44,45) was selected for enthalpy of formation of ideal gas. Data for enthalpy of formation of the ideal gas is a series expansion in temperature, Equation (1-11). Results from the correlation are in favorable agreement with data.

GIBB'S FREE ENERGY OF FORMATION - Figure 7-12

Results from the data compilation of Yaws and co-workers (44,46) were selected for Gibb's free energy of formation of ideal gas. Data for Gibb's free energy of formation of the ideal gas is a series expansion in temperature, Equation (1-12). Correlation results are in favorable agreement with data.

Table 7-1 Physical Properties

	Ethyl Chloride	Propyl Chloride	Ethylene Dichloride	Propylene Dichloride
1. Name	Ethyl Chloride	Propyl Chloride	Ethylene Dichloride	Propylene Dichloride
2. Formula	C_2H_5Cl	C_3H_7Cl	$C_2H_4Cl_2$	$C_3H_6Cl_2$
3. Molecular Weight, g/mol	64.514	78.541	98.960	112.986
4. Critical Temperature, K	460.35	503.10	561.00	572.00
5. Critical Pressure, bar	52.689	45.8	53.702	42.400
6. Critical Volume, ml/mol	200.00	254.00	220.00	291.00
7. Critical Compressibility Factor	0.275	0.278	0.253	0.259
8. Acentric Factor	0.1905	0.2350	0.2876	0.2513
9. Melting Point, K	136.75	150.40	237.49	172.71
10. Boiling Point @ 1 atm, K	285.42	319.70	356.59	369.52
11. Heat of Vaporization @ BP, kJ/kg	382.712	346.800	323.751	277.723
12. Density of Liquid @ 25 C, g/ml	0.888	0.885	1.233	1.139
13. Surface Tension @ 25 C, dynes/cm	18.53	21.29	31.84	28.31
14. Heat Capacity of Gas @ 25 C, J/g K	0.97	1.08	0.80	0.87
15. Heat Capacity of Liquid @ 25 C, J/g K	1.66	1.69	1.26	1.36
16. Viscosity of Gas @ 25 C, micropoise	97.23	77.00	92.03	83.03
17. Viscosity of Liquid @ 25 C, centipoise	------	0.3475	0.7698	0.7561
18. Thermal Conductivity of Gas @ 25 C, W/m K	0.0114	0.0076	0.0086	0.0069
19. Thermal Conductivity of Liquid @ 25 C, W/m K	0.1191	0.1540	0.1347	0.1234
20. Enthalpy of Formation of Gas @ 25 C, kJ/mol	-111.78	-130.22	-129.73	-165.77
21. Gibbs Free Energy of Formation of Gas @ 25 C, kJ/mol	-60.18	-50.91	-73.98	-83.31
22. Flash Point, K	223.15	255.45	286.48	288.71
23. Autoignition Temperature, K	792.04	------	685.93	830.37
24. Lower Explosion Limit in Air, vol %	3.8	2.5	6.2	3.4
25. Upper Explosion Limit in Air, vol %	15.4	11.0	16.0	14.5
26. Solubility in Water @ 25 C, ppm(wt)	9051 @ 20 C	2700 @ 20 C	8679	2750

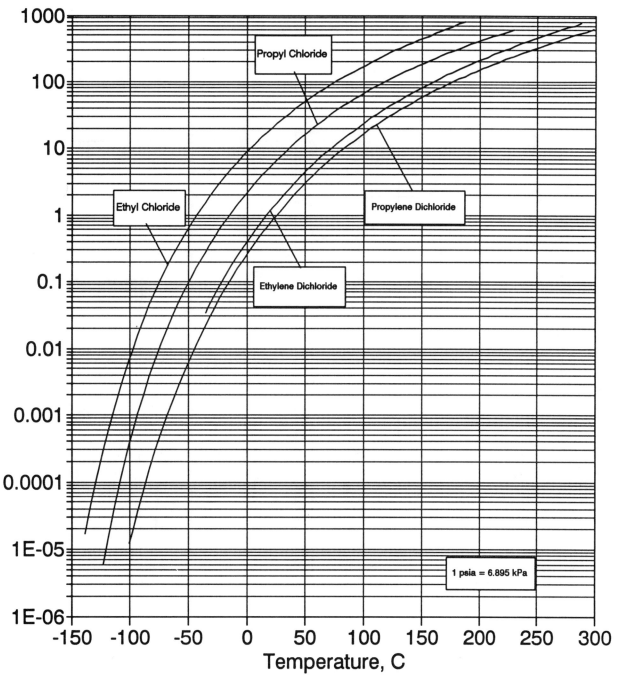

Figure 7-1 Vapor Pressure

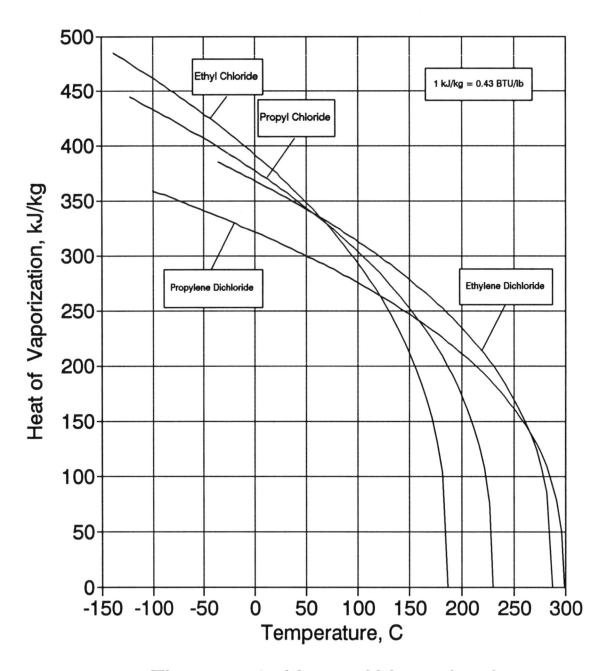

Figure 7-2 Heat of Vaporization

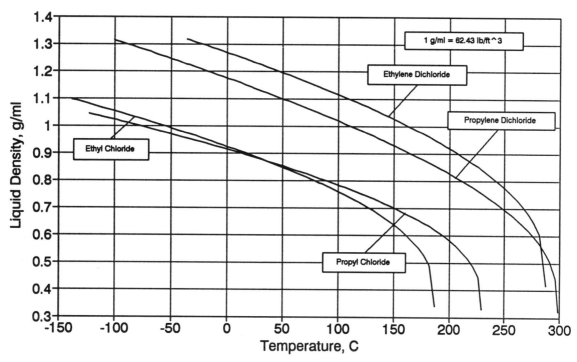

Figure 7-3 Liquid Density

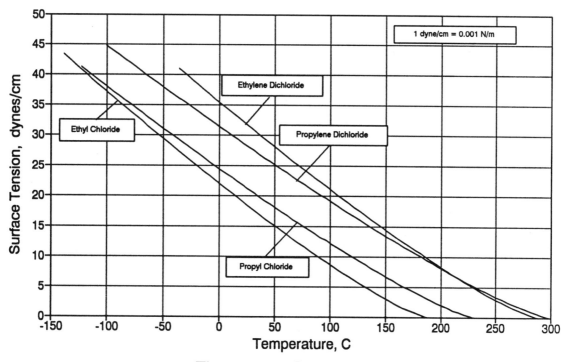

Figure 7-4 Surface Tension

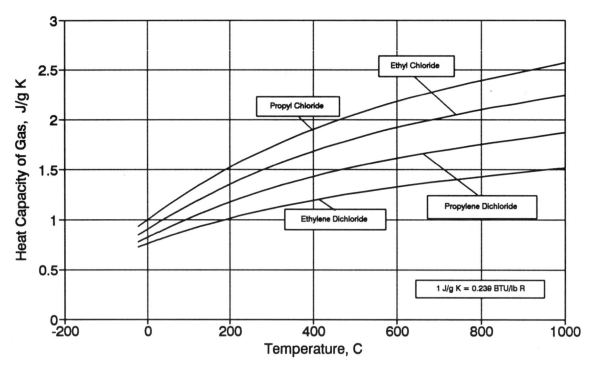

Figure 7-5 Heat Capacity of Gas

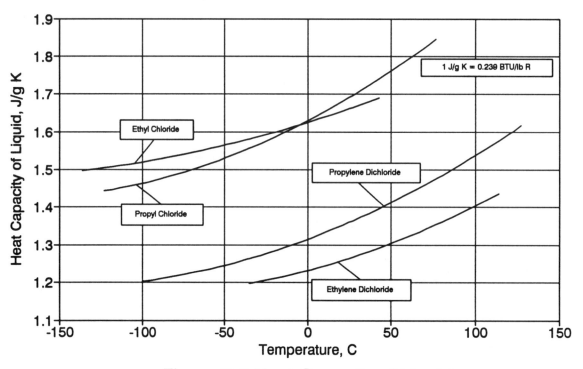

Figure 7-6 Heat Capacity of Liquid

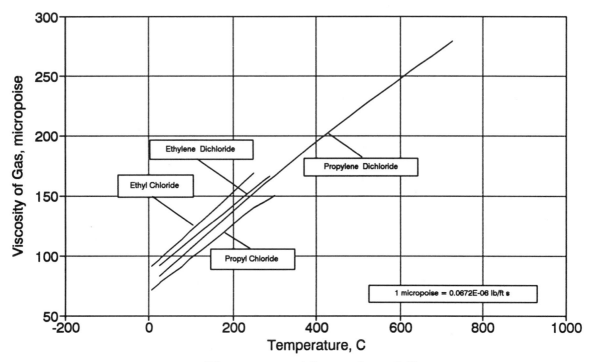

Figure 7-7 Viscosity of Gas

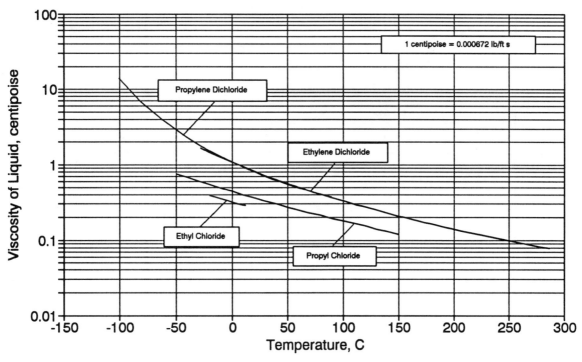

Figure 7-8 Viscosity of Liquid

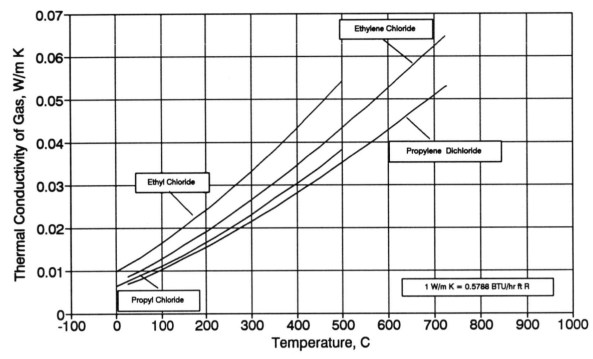

Figure 7-9 Thermal Conductivity of Gas

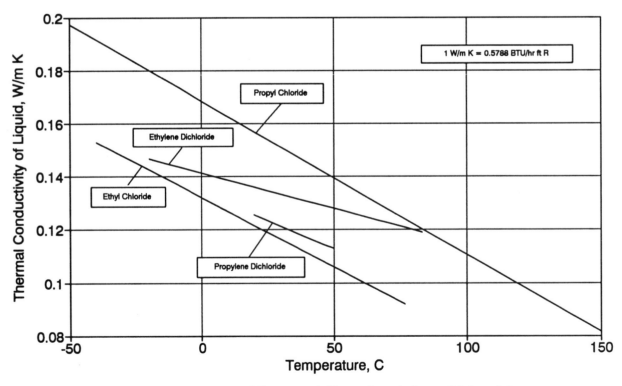

Figure 7-10 Thermal Conductivity of Liquid

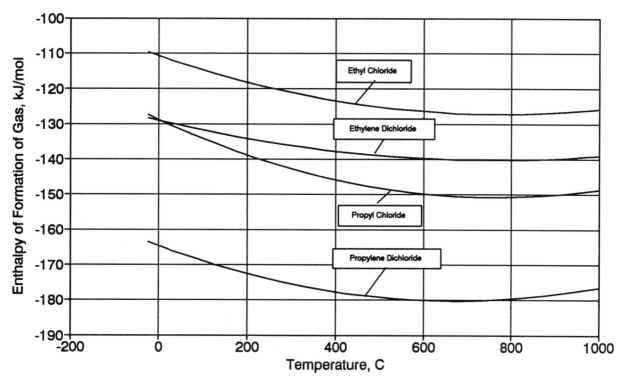

Figure 7-11 Enthalpy of Formation of Gas

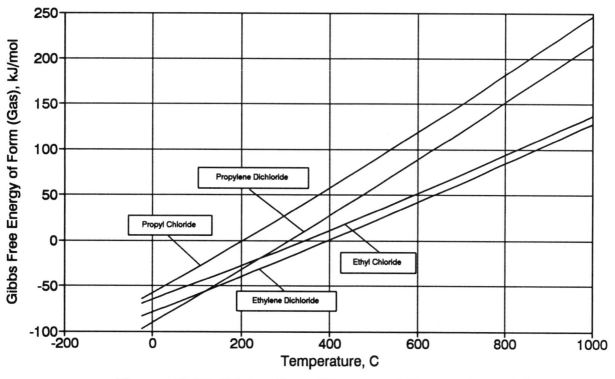

Figure 7-12 Gibbs Free Energy of Formation of Gas

Chapter 8

PRIMARY ALCOHOLS

Robert W. Gallant and Carl L. Yaws

PHYSICAL PROPERTIES - Table 8-1

Property data from the literature (1-55,110,112-115,126-137) are given in Table 8-1. The critical constants have been determined experimentally for the primary alcohols (1-7). Additional property data such as acentric factor, enthalpy of formation, lower explosion limit in air and solubility in water are also available. The DIPPR (Design Institute for Physical Property Research) project (5) and recent data compilations by Yaws and co-workers (44-55) were consulted extensively in preparing the tabulation.

VAPOR PRESSURE - Figure 8-1

Results from the DIPPR project (5) were selected for vapor pressure from very low temperatures to the critical point. Correlation of data for vapor pressure as a function of temperature was accomplished using Equation (1-1). Results from this equation (Antoine-type with extended terms) are in favorable agreement with experimental data. Errors are about 1-5% or less in most cases.

HEAT OF VAPORIZATION - Figure 8-2

The data compilation of Yaws and co-workers (44,52) was selected for heat of vaporization for temperatures ranging from melting point to critical point. The Watson equation, Equation (1-2), was used for correlation of the data as a function of temperature. Reliability of results is good with errors of about 1-5% or less.

LIQUID DENSITY - Figure 8-3

Results from the data compilation of Yaws and co-workers (44,54) were selected for liquid density from low temperatures at the melting point to higher temperatures up to the critical point. A modified Rackett equation, Equation (1-3), was used for correlation of the data as a function of temperature. Results from the correlation are in favorable agreement with data. Deviations are less than 1-2% in most cases.

SURFACE TENSION - Figure 8-4

The data compilation of Yaws and co-workers (44,55) was selected for surface tension for temperatures from melting point to critical point. Using data from the literature, correlation for surface tension as a function of temperature over the full liquid range was achieved by the modified Othmer equation, Equation (1-4). Accuracy is good with errors being about 1-10% or less in most cases.

HEAT CAPACITY - Figures 8-5 and 8-6

Results from the data compilation of Yaws and co-workers (44) were selected for heat capacity of ideal gas. Correlation of data was accomplished using a series expansion in temperature, Equation (1-5). Results are in favorable agreement with data. Errors are about 1% or less in most cases.

Results from the DIPPR project (5) were selected for heat capacity of liquid. The coverage applies to temperatures from below the boiling point to temperatures above the boiling point for most of the compounds. Data were correlated with a series expansion in temperature, Equation (1-6). Results are in favorable agreement with data. Errors are about 5% or less using the correlation.

VISCOSITY - Figures 8-7 and 8-8

The DIPPR project (5) was selected for viscosity of gas. Data for gas viscosity as a function of temperature were correlated using Equation (1-7). Results are in agreement with data. Errors are about 10% or less in most cases.

The DIPPR project (5) was also selected for viscosity of liquid. Temperatures from below the boiling point to temperatures above the boiling point are covered for most of the compounds. Data for liquid viscosity as a function of temperature were correlated using the de Guzman - Andrade equation with extended terms, Equation (1-8). Correlation results and data are in agreement with errors being about 10% or less.

THERMAL CONDUCTIVITY - Figures 8-9 and 8-10

Results from the DIPPR project (5) were selected for thermal conductivity of gas. Data for gas thermal conductivity as a function of temperature were correlated using the Equation (1-9). Reliability of results is good with errors of about 3-10% or less in most cases.

Results from the DIPPR project (5) were selected for thermal conductivity of liquid. The coverage applies to temperatures from below the boiling point to temperatures above the boiling point for most of the compounds. Data for liquid thermal conductivity as a function of temperature were correlated using a series expansion in temperature, Equation (1-10). Results are in favorable agreement with data. Errors are about 5% or less in most cases.

ENTHALPY OF FORMATION - Figure 8-11

The data compilation of Yaws and co-workers (44,45) was selected for enthalpy of formation of ideal gas. Data for enthalpy of formation of the ideal gas is a series expansion in temperature, Equation (1-11). Results from the correlation are in favorable agreement with data.

GIBB'S FREE ENERGY OF FORMATION - Figure 8-12

Results from the data compilation of Yaws and co-workers (44,46) were selected for Gibb's free energy of formation of ideal gas. Data for Gibb's free energy of formation of the ideal gas is a series expansion in temperature, Equation (1-12). Correlation results are in favorable agreement with data.

Table 8-1 Physical Properties

	Methanol	Ethanol	Propanol	Butanol
1. Name	Methanol	Ethanol	Propanol	Butanol
2. Formula	CH_4O	C_2H_6O	C_3H_8O	$C_4H_{10}O$
3. Molecular Weight, g/mol	32.042	46.069	60.096	74.122
4. Critical Temperature, K	512.58	516.25	536.71	562.93
5. Critical Pressure, bar	80.959	63.835	51.696	44.127
6. Critical Volume, ml/mol	117.80	166.92	218.53	274.53
7. Critical Compressibility Factor	0.224	0.248	0.253	0.259
8. Acentric Factor	0.5656	0.6371	0.6279	0.5945
9. Melting Point, K	175.47	159.05	146.95	183.85
10. Boiling Point @ 1 atm, K	337.85	351.44	370.35	390.81
11. Heat of Vaporization @ BP, kJ/kg	1100.1	841.12	694.90	581.14
12. Density of Liquid @ 25 C, g/ml	0.785	0.783	0.798	0.837
13. Surface Tension @ 25 C, dynes/cm	21.88	21.78	23.12	24.82
14. Heat Capacity of Gas @ 25 C, J/g K	1.367	1.419	1.448	1.482
15. Heat Capacity of Liquid @ 25 C, J/g K	2.533	2.434	2.407	2.393
16. Viscosity of Gas @ 25 C, micropoise	96.14	89.34	75.75	69.30
17. Viscosity of Liquid @ 25 C, centipoise	0.549	1.082	1.965	2.637
18. Thermal Conductivity of Gas @ 25 C, W/m K	0.0156	------	------	------
19. Thermal Conductivity of Liquid @ 25 C, W/m K	0.2026	0.1692	0.1536	0.1536
20. Enthalpy of Formation of Gas @ 25 C, kJ/mol	-201.20	-234.88	-257.61	-274.55
21. Gibbs Free Energy of Formation of Gas @ 25 C, kJ/mol	-162.66	-168.51	-163.26	-151.04
22. Flash Point, K	284.26	285.93	288.15	302.04
23. Autoignition Temperature, K	737.04	695.93	644.26	616.48
24. Lower Explosion Limit in Air, vol %	7.3	4.3	2.0	1.4
25. Upper Explosion Limit in Air, vol %	36.0	19.0	12.0	11.2
26. Solubility in Water @ 25 C, ppm(wt)	total	total	total	74600

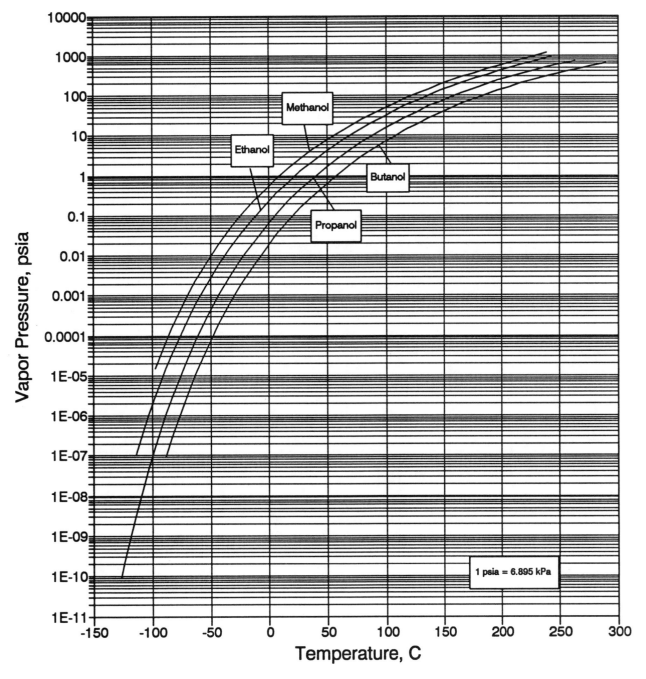

Figure 8-1 Vapor Pressure

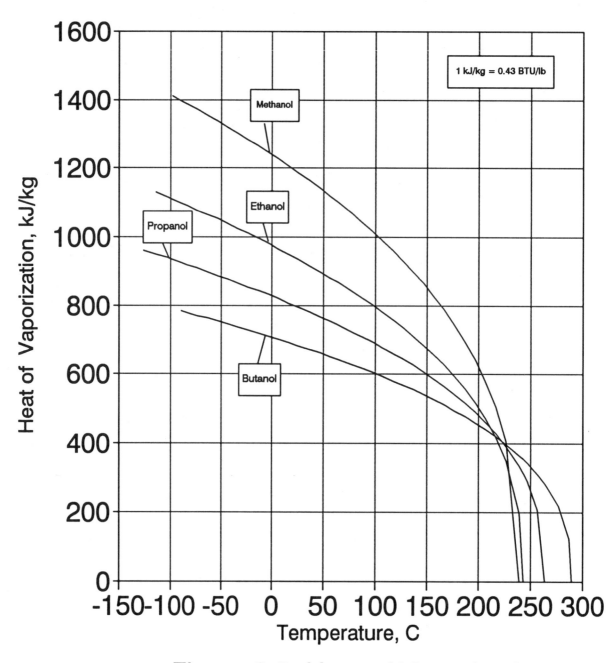

Figure 8-2 Heat of Vaporization

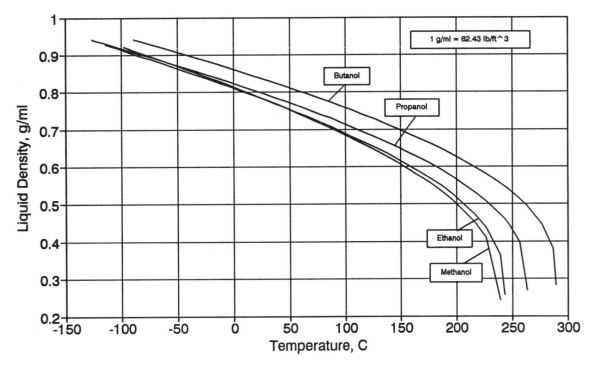

Figure 8-3 Liquid Density

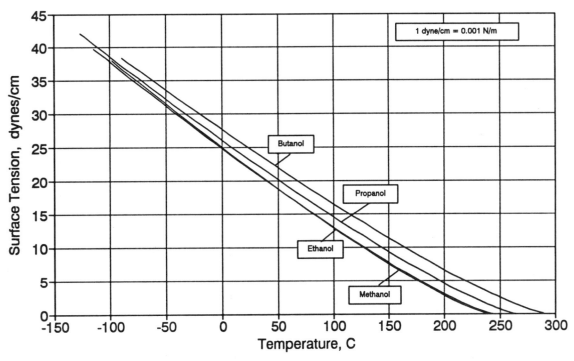

Figure 8-4 Surface Tension

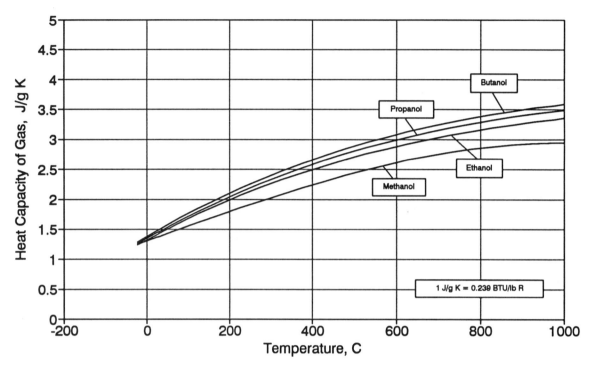

Figure 8-5 Heat Capacity of Gas

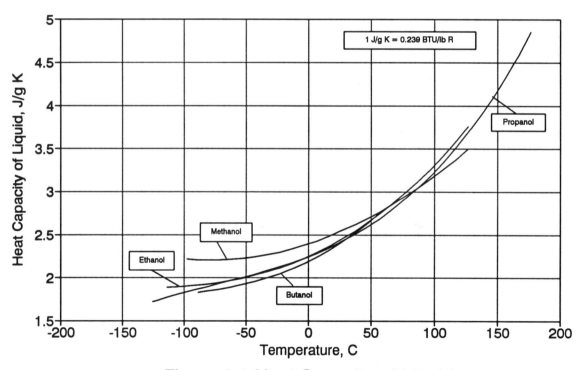

Figure 8-6 Heat Capacity of Liquid

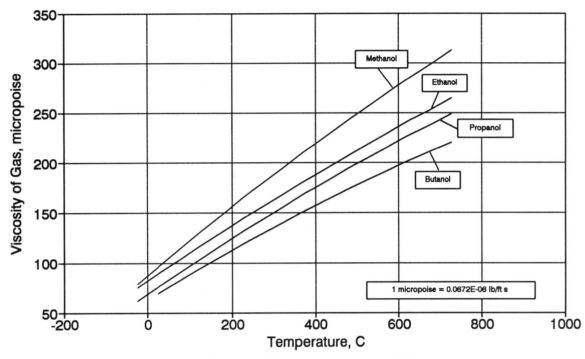

Figure 8-7 Viscosity of Gas

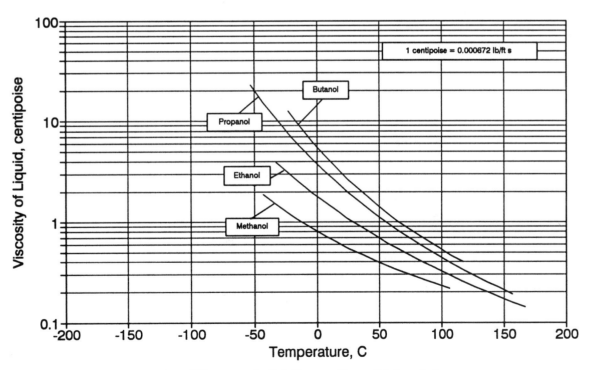

Figure 8-8 Viscosity of Liquid

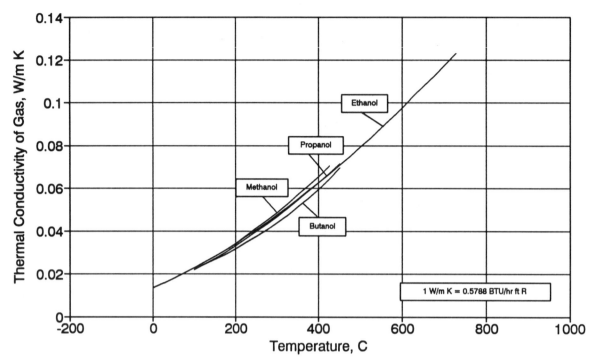

Figure 8-9 Thermal Conductivity of Gas

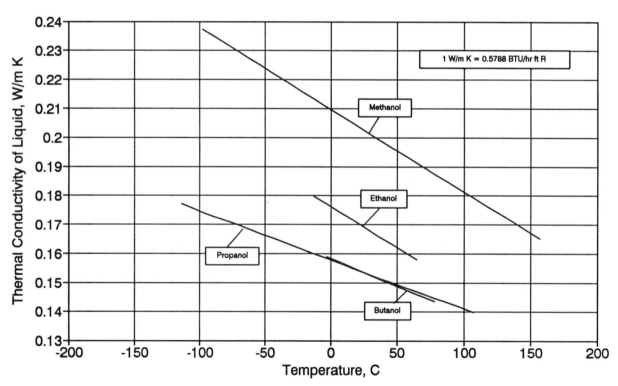

Figure 8-10 Thermal Conductivity of Liquid

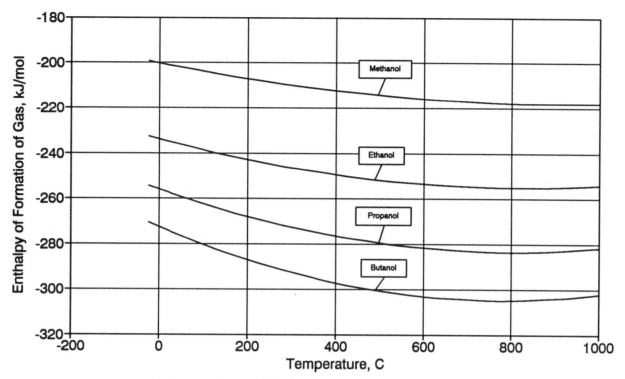

Figure 8-11 Enthalpy of Formation of Gas

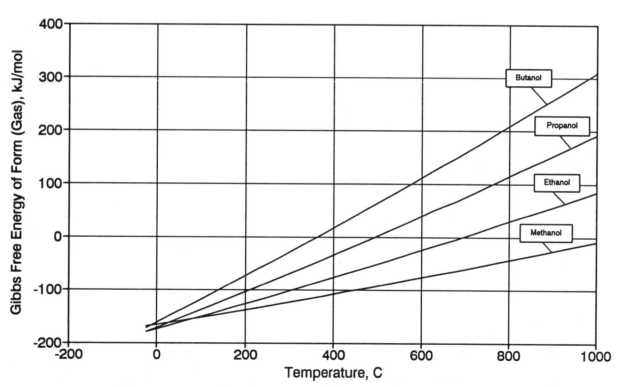

Figure 8-12 Gibbs Free Energy of Formation of Gas

Chapter 9

C$_3$ TO C$_4$ ALCOHOLS

Robert W. Gallant, Carl L. Yaws and Xiang Pan

PHYSICAL PROPERTIES - Table 9-1

Property data from the literature (1-55,110-115,126,131,135,147-164) are given in Table 9-1. The critical constants have been determined experimentally for the alcohols (1-7). Additional property data such as acentric factor, enthalpy of formation, lower explosion limit in air and solubility in water are also available. The DIPPR (Design Institute for Physical Property Research) project (5) and recent data compilations by Yaws and co-workers (44-55) were consulted extensively in preparing the tabulation.

VAPOR PRESSURE - Figure 9-1

Results from the DIPPR project (5) were selected for vapor pressure from very low temperatures to the critical point. Correlation of data for vapor pressure as a function of temperature was accomplished using Equation (1-1). Results from this equation (Antoine-type with extended terms) are in favorable agreement with experimental data. Errors are about 1-5% or less in most cases.

HEAT OF VAPORIZATION - Figure 9-2

For all compounds except isobutanol (5), the data compilation of Yaws and co-workers (44,52) was selected for heat of vaporization for temperatures ranging from melting point to critical point. The Watson equation, Equation (1-2), was used for correlation of the data as a function of temperature. Reliability of results is good with errors of about 1-5% or less.

LIQUID DENSITY - Figure 9-3

For all compounds except isobutanol (5), results from the data compilation of Yaws and co-workers (44,54) were selected for liquid density from low temperatures at the melting point to higher temperatures up to the critical point. A modified Rackett equation, Equation (1-3), was used for correlation of the data as a function of temperature. Results from the correlation are in favorable agreement with data. Deviations are less than 1-2% in most cases.

SURFACE TENSION - Figure 9-4

For all compounds except isobutanol (5), the data compilation of Yaws and co-workers (44,55) was selected for surface tension for temperatures from melting point to critical point. Using data from the literature, correlation for surface tension as a function of temperature over the full liquid range was achieved by the modified Othmer equation, Equation (1-4). Accuracy is good with errors being about 1-10% or less in most cases.

HEAT CAPACITY - Figures 9-5 and 9-6

For all compounds except isobutanol (5), results from the data compilation of Yaws and co-workers (44) were selected for heat capacity of ideal gas. Correlation of data was accomplished using a series expansion in temperature, Equation (1-5). Results are in favorable agreement with data. Errors are about 1% or less in most cases.

Results from the DIPPR project (5) were selected for heat capacity of liquid. The coverage applies to temperatures from below the boiling point to temperatures above the boiling point for most of the compounds. Data were correlated with a series expansion in temperature, Equation (1-6). Results are in favorable agreement with data. Errors are about 5% or less using the correlation.

VISCOSITY - Figures 9-7 and 9-8

The DIPPR project (5) was selected for viscosity of gas. Data for gas viscosity as a function of temperature were correlated using Equation (1-7). Results are in agreement with data. Errors are about 5-10% or less for all compounds except isobutanol (possible 25-50% error range, 5).

The DIPPR project (5) was also selected for viscosity of liquid. Temperatures from below the boiling point to temperatures above the boiling point are covered for most of the compounds. Data for liquid viscosity as a function of temperature were correlated using the de Guzman - Andrade equation with extended terms, Equation (1-8). Correlation results and data are in agreement with errors being about 3-10% or less.

THERMAL CONDUCTIVITY - Figures 9-9 and 9-10

Results from the DIPPR project (5) were selected for thermal conductivity of gas. Data for gas thermal conductivity as a function of temperature were correlated using the Equation (1-9). Reliability of results is fair with errors of about 3-25% or less in most cases.

Results from the DIPPR project (5) were selected for thermal conductivity of liquid. The coverage applies to temperatures from below the boiling point to temperatures above the boiling point for most of the compounds. Data for liquid thermal conductivity as a function of temperature were correlated using a series expansion in temperature, Equation (1-10). Results are in favorable agreement with data. Errors are about 5-10% or less in most cases.

ENTHALPY OF FORMATION - Figure 9-11

For all compounds except isobutanol (5), the data compilation of Yaws and co-workers (44,45) was selected for enthalpy of formation of ideal gas. For isobutanol, the variation of values with temperature was estimated from the temperature behavior of values for t-butanol and s-butanol. Data for enthalpy of formation of the ideal gas is a series expansion in temperature, Equation (1-11). Results from the correlation are in favorable agreement with data.

GIBB'S FREE ENERGY OF FORMATION - Figure 9-12

For all compounds except isobutanol (5), results from the data compilation of Yaws and co-workers (44,46) were selected for Gibb's free energy of formation of ideal gas. For isobutanol, the variation of values with temperature was estimated from the temperature behavior of values for t-butanol and s-butanol. Data for Gibb's free energy of formation of the ideal gas is a series expansion in temperature, Equation (1-12). Correlation results are in favorable agreement with data.

Table 9-1 Physical Properties

	Isopropanol	T-Butanol	S-Butanol	Isobutanol
1. Name	Isopropanol	T-Butanol	S-Butanol	Isobutanol
2. Formula	C3H8O	C4H10O	C4H10O	C4H10O
3. Molecular Weight, g/mol	60.096	74.122	74.122	74.122
4. Critical Temperature, K	508.31	506.20	536.01	547.73
5. Critical Pressure, bar	47.643	39.719	41.938	42.952
6. Critical Volume, ml/mol	220.13	275.00	268.00	272.00
7. Critical Compressibility Factor	0.248	0.260	0.252	0.257
8. Acentric Factor	0.6689	0.6158	0.5711	0.5885
9. Melting Point, K	184.65	298.81	158.45	165.15
10. Boiling Point @ 1 atm, K	355.41	355.57	372.70	380.81
11. Heat of Vaporization @ BP, kJ/kg	662.95	527.10	550.49	586.42
12. Density of Liquid @ 25 C, g/ml	0.775	------	0.801	0.797
13. Surface Tension @ 25 C, dynes/cm	20.72	------	22.37	22.56
14. Heat Capacity of Gas @ 25 C, J/g K	1.47	1.53	1.53	1.48
15. Heat Capacity of Liquid @ 25 C, J/g K	2.60	------	2.65	2.45
16. Viscosity of Gas @ 25 C, micropoise	76.91	67.07	73.94	71.17
17. Viscosity of Liquid @ 25 C, centipoise	2.055	------	3.174	3.334
18. Thermal Conductivity of Gas @ 25 C, W/m K	0.0197	0.0158	0.0133	0.0116
19. Thermal Conductivity of Liquid @ 25 C, W/m K	0.1400	------	0.1342	0.1399
20. Enthalpy of Formation of Gas @ 25 C, kJ/mol	-272.73	-325.97	-292.43	-283.22
21. Gibbs Free Energy of Formation of Gas @ 25 C, kJ/mol	-173.88	-191.40	-167.66	-155.01
22. Flash Point, K	284.82	284.26	297.04	300.93
23. Autoignition Temperature, K	672.04	750.93	679.26	699.82
24. Lower Explosion Limit in Air, vol %	2.0	2.4	1.7	1.7
25. Upper Explosion Limit in Air, vol %	12.0	8.0	9.8	10.9
26. Solubility in Water @ 25 C, ppm(wt)	total	total	51800	75000 @ 30 C

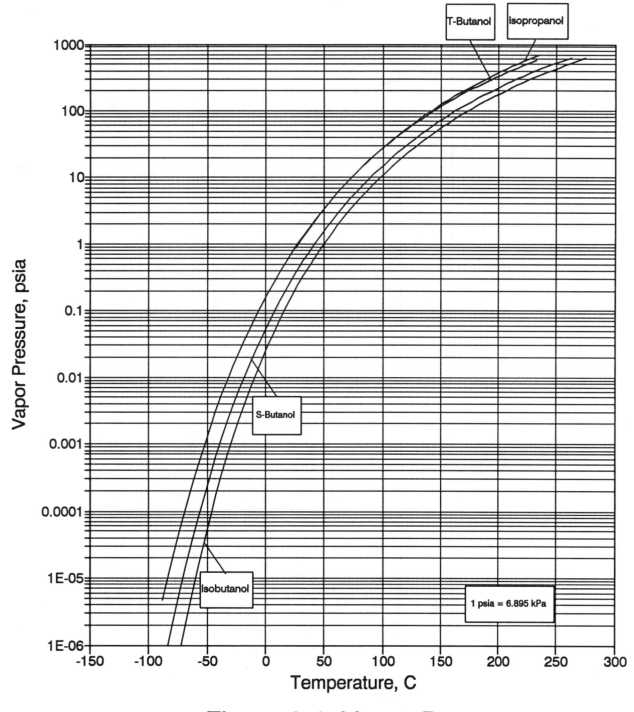

Figure 9-1 Vapor Pressure

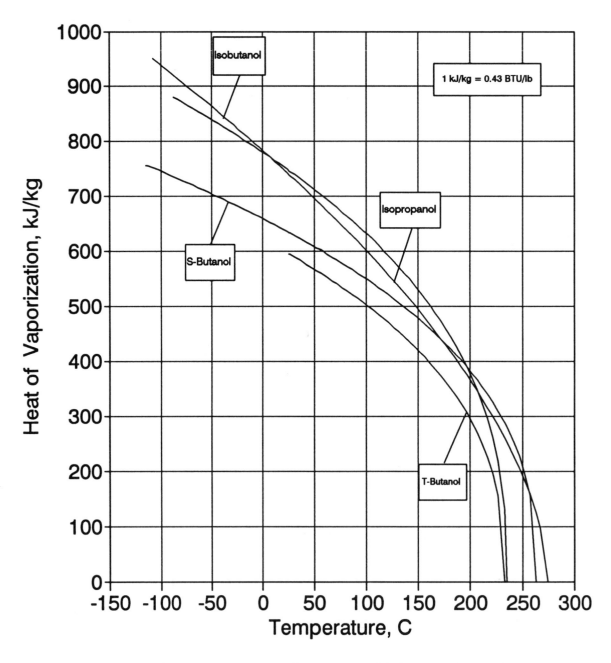

Figure 9-2 Heat of Vaporization

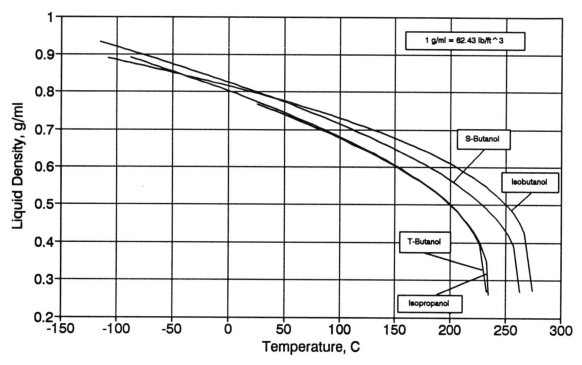

Figure 9-3 Liquid Density

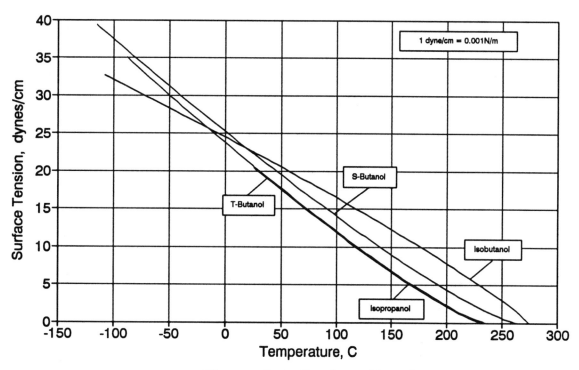

Figure 9-4 Surface Tension

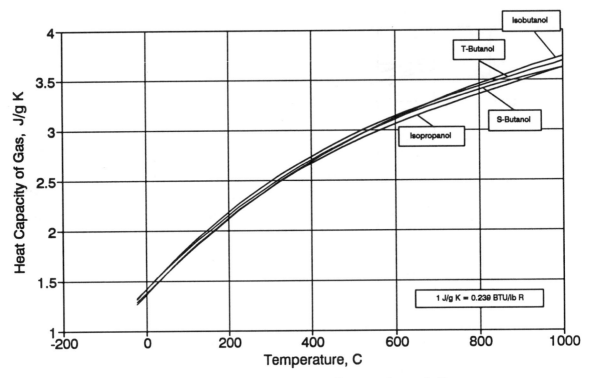

Figure 9-5 Heat Capacity of Gas

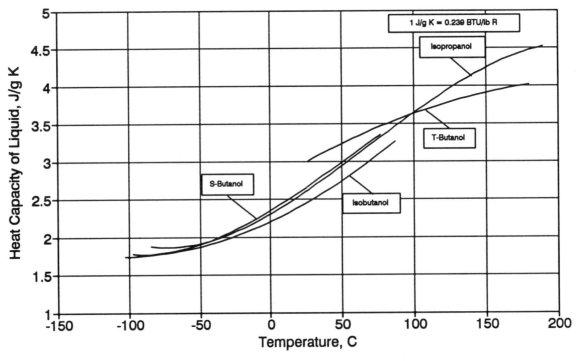

Figure 9-6 Heat Capacity of Liquid

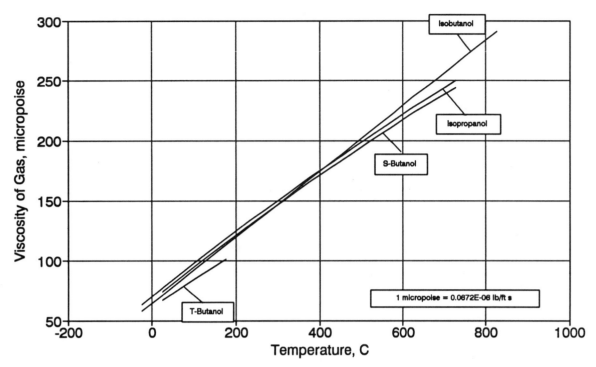

Figure 9-7 Viscosity of Gas

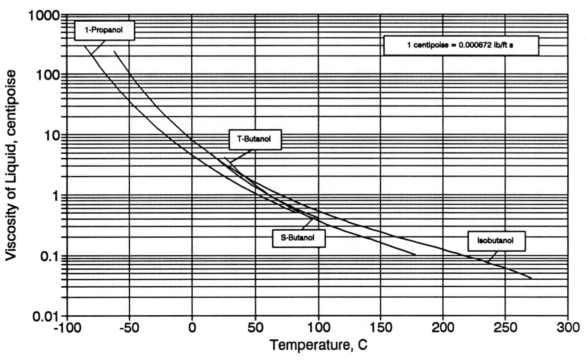

Figure 9-8 Viscosity of Liquid

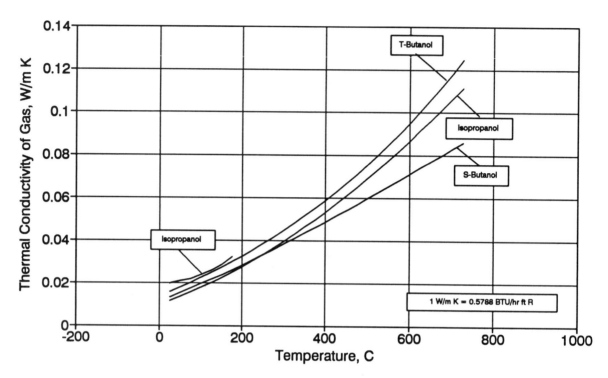

Figure 9-9 Thermal Conductivity of Gas

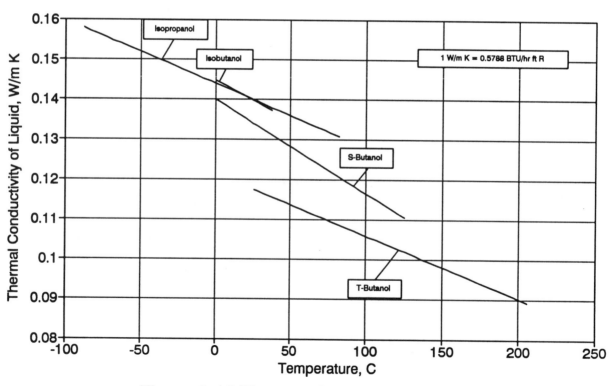

Figure 9-10 Thermal Conductivity of Liquid

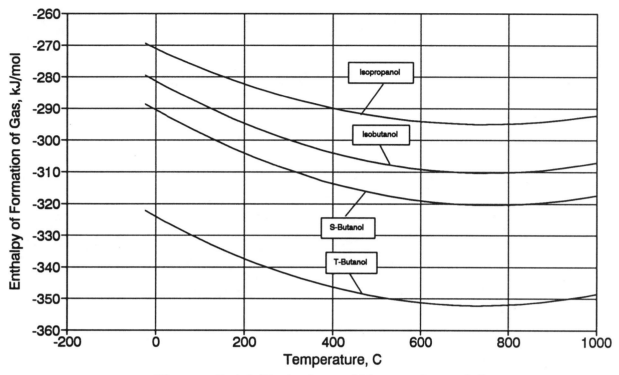

Figure 9-11 Enthalpy of Formation of Gas

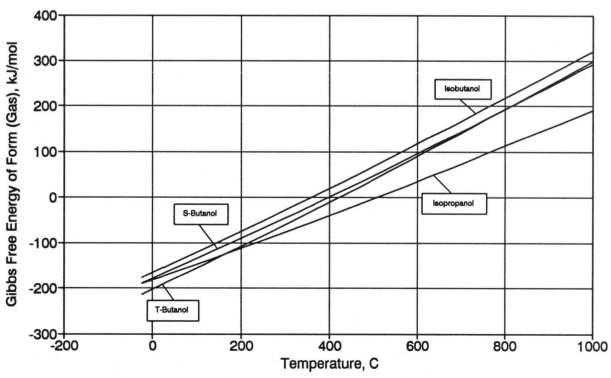

Figure 9-12 Gibbs Free Energy of Formation of Gas

Chapter 10

MISCELLANEOUS ALCOHOLS

Robert W. Gallant and Carl L. Yaws

PHYSICAL PROPERTIES - Table 10-1

Property data from the literature (1-55,110,113,114,126,131,136,138-146) are given in Table 10-1. Since results for allyl alcohol and 1-heptanol are not available in the DIPPR project, critical constants for these compounds were selected from Yaws (44,47). Critical constants for the remaining compounds were selected from the DIPPR project (5). Additional property data such as acentric factor, enthalpy of formation, lower explosion limit in air and solubility in water are also available. The DIPPR (Design Institute for Physical Property Research) project (5) and recent data compilations by Yaws and co-workers (44-55) were consulted extensively in preparing the tabulation.

VAPOR PRESSURE - Figure 10-1

For allyl alcohol and 1-heptanol, results were selected from Yaws and co-workers (44,48). For the remaining compounds, results from the DIPPR project (5) were selected for vapor pressure from very low temperatures to the critical point. Correlation of data for vapor pressure as a function of temperature was accomplished using Equation (1-1). Results from this equation (Antoine-type with extended terms) are in favorable agreement with experimental data. Errors are about 1-5% or less in most cases.

HEAT OF VAPORIZATION - Figure 10-2

The data compilation of Yaws and co-workers (44,52) was selected for heat of vaporization for temperatures ranging from melting point to critical point. The Watson equation, Equation (1-2), was used for correlation of the data as a function of temperature. Reliability of results is good with errors of about 5% or less in most cases.

LIQUID DENSITY - Figure 10-3

Results from the data compilation of Yaws and co-workers (44,54) were selected for liquid density from low temperatures at the melting point to higher temperatures up to the critical point. A modified Rackett equation, Equation (1-3), was used for correlation of the data as a function of temperature. Results from the correlation are in favorable agreement with data. Deviations are less than 2% in most cases.

SURFACE TENSION - Figure 10-4

For 1-heptanol, the original data of Mueller (145) as reported by Gallant (30) was used. For the remaining compounds, the data compilation of Yaws and co-workers (44,55) was selected for surface tension for temperatures from melting point to critical point. Using data from the literature, correlation for surface tension as a function of temperature over the full liquid range was achieved by the modified Othmer equation, Equation (1-4). Accuracy is good with errors being about 1-10% or less in most cases.

HEAT CAPACITY - Figures 10-5 and 10-6

Results from the data compilation of Yaws and co-workers (44) were selected for heat capacity of ideal gas. Correlation of data was accomplished using a series expansion in temperature, Equation (1-5). Results are in favorable agreement with data. Errors are about 1% or less in most cases.

Results from Yaws and co-workers (44) were selected for liquid heat capacity. The coverage applies to temperatures from below the boiling point to temperatures above the boiling point for most of the compounds. Data were correlated with a series expansion in temperature, Equation (1-6). Correlation results are in favorable agreement with data.

VISCOSITY - Figures 10-7 and 10-8

Since results for allyl alcohol and 1-heptanol are not available in the DIPPR project, values for gas and liquid viscosities were estimated (30). Since these compounds are polar and data are not available, the estimates should be considered rough approximations.

For the remaining compounds, results from the DIPPR project (5) were selected for viscosity of gas. Data for gas viscosity as a function of temperature were correlated using Equation (1-7). Results are in favorable agreement with data. Errors are about 10% or less in most cases.

For the remaining compounds, results from the DIPPR project (5) were selected for viscosity of liquid. Temperatures from below the boiling point to temperatures above the boiling point are covered for most of the compounds. Data for liquid viscosity as a function of temperature were correlated using the de Guzman - Andrade equation with extended terms, Equation (1-8). Correlation results and data are in favorable agreement with errors being about 3-5% or less.

THERMAL CONDUCTIVITY - Figures 10-9 and 10-10

Since results for allyl alcohol and 1-heptanol are not available in the DIPPR project, values for gas and liquid thermal conductivities were estimated (30). Since these compounds are polar and data are not available, the estimates should be considered rough approximations.

Results from the DIPPR project (5) were selected for thermal conductivity of gas for the remaining compounds. Data for gas thermal conductivity as a function of temperature were correlated using the Equation (1-9). Reliability of results is good with errors of about 10% or less in most cases.

Results from the DIPPR project (5) were selected for liquid thermal conductivity of the remaining compounds. The coverage applies to temperatures from below the boiling point to temperatures above the boiling point for most of the compounds. Data for liquid thermal conductivity as a function of temperature were correlated using a series expansion in temperature, Equation (1-10). Results are in fair agreement with data. Errors are about 5% or less in most cases.

ENTHALPY OF FORMATION - Figure 10-11

The data compilation of Yaws and co-workers (44,45) was selected for enthalpy of formation of ideal gas. Data for enthalpy of formation of the ideal gas is a series expansion in temperature, Equation (1-11). Results from the correlation are in favorable agreement with data.

GIBB'S FREE ENERGY OF FORMATION - Figure 10-12

Results from the data compilation of Yaws and co-workers (44,46) were selected for Gibb's free energy of formation of ideal gas. Data for Gibb's free energy of formation of the ideal gas is a series expansion in temperature, Equation (1-12). Correlation results are in favorable agreement with data.

Table 10-1 Physical Properties

1. Name	Allyl Alcohol	1-Pentanol	1-Hexanol	1-Heptanol
2. Formula	C3H6O	C5H12O	C6H14O	C7H16O
3. Molecular Weight, g/mol	58.080	88.149	102.176	116.203
4. Critical Temperature, K	545.0	586.15	611.35	631.9
5. Critical Pressure, bar	53.1	38.8	35.1	31.6
6. Critical Volume, ml/mol	203.5	326.0	381.26	435.8
7. Critical Compressibility Factor	0.240	0.260	0.263	0.262
8. Acentric Factor	0.554	0.5938	0.5803	0.560
9. Melting Point, K	144.0	194.95	228.55	239.2
10. Boiling Point @ 1 atm, K	369.7	410.95	430.15	449.1
11. Heat of Vaporization @ BP, kJ/kg	687.96	503.07	475.15	414.07
12. Density of Liquid @ 25 C, g/ml	0.844	0.811	0.815	0.818
13. Surface Tension @ 25 C, dynes/cm	25.11	25.26	25.32	25.34
14. Heat Capacity of Gas @ 25 C, J/g K	1.308	1.506	1.523	1.536
15. Heat Capacity of Liquid @ 25 C, J/g K	2.08	2.22	2.28	1.98
16. Viscosity of Gas @ 25 C, micropoise	79.0	-----	-----	63.0
17. Viscosity of Liquid @ 25 C, centipoise	1.25	3.484	4.578	5.800
18. Thermal Conductivity of Gas @ 25 C, W/m K	0.0167	------	------	0.0096
19. Thermal Conductivity of Liquid @ 25 C, W/m K	0.164	0.153	0.150	0.140
20. Enthalpy of Formation of Gas @ 25 C, kJ/mol	-132.08	-302.53	-319.78	-335.04
21. Gibbs Free Energy of Formation of Gas @ 25 C, kJ/mol	-71.46	-150.15	-138.45	-124.73
22. Flash Point, K	294.15	305.93	335.93	349.75
23. Autoignition Temperature, K	648.15	573.15	665.15	------
24. Lower Explosion Limit in Air, vol %	------	1.2	------	------
25. Upper Explosion Limit in Air, vol %	------	10.0	------	------
26. Solubility in Water @ 25 C, ppm(wt)	total	22000	5875	1740

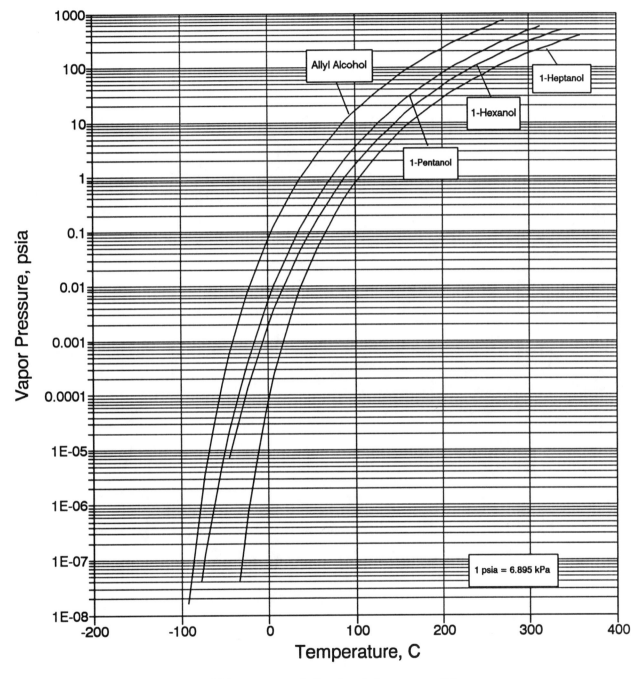

Figure 10-1 Vapor Pressure

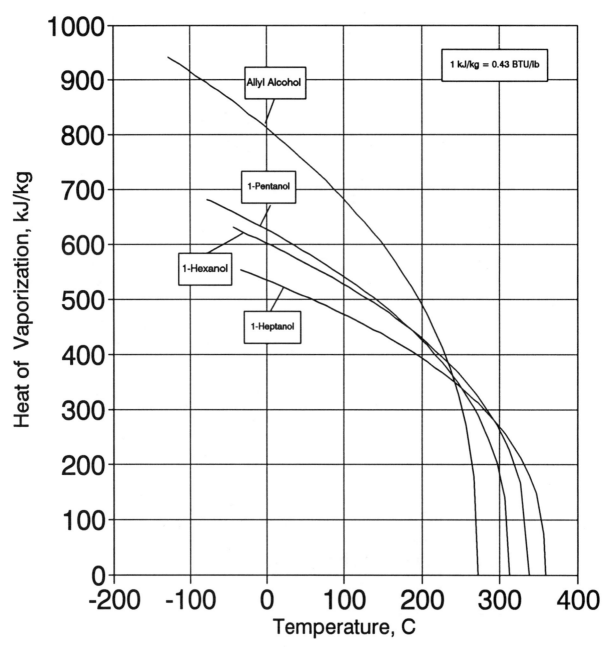

Figure 10-2 Heat of Vaporization

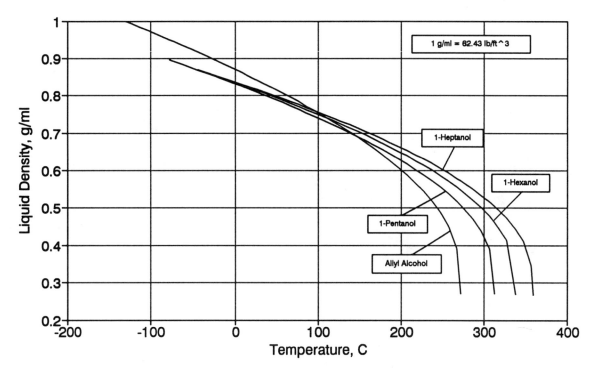

Figure 10-3 Liquid Density

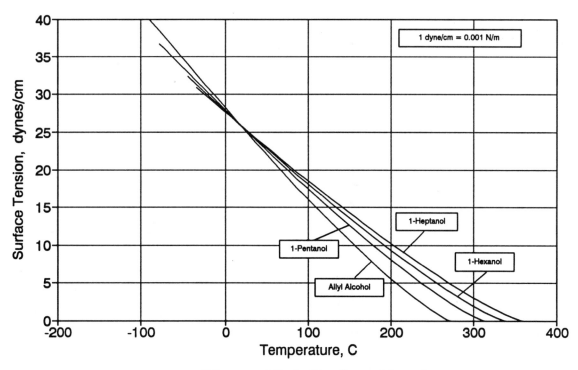

Figure 10-4 Surface Tension

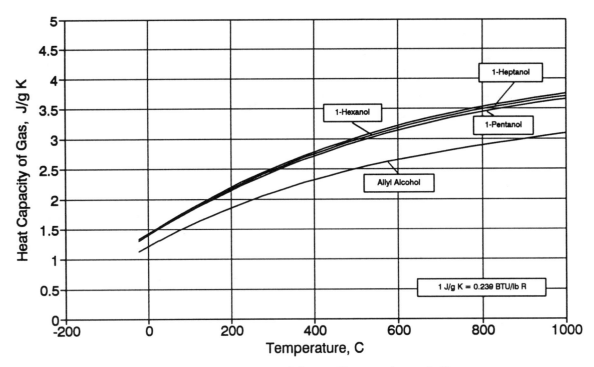

Figure 10-5 Heat Capacity of Gas

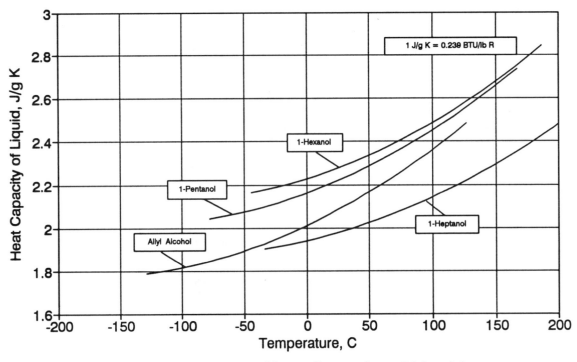

Figure 10-6 Heat Capacity of Liquid

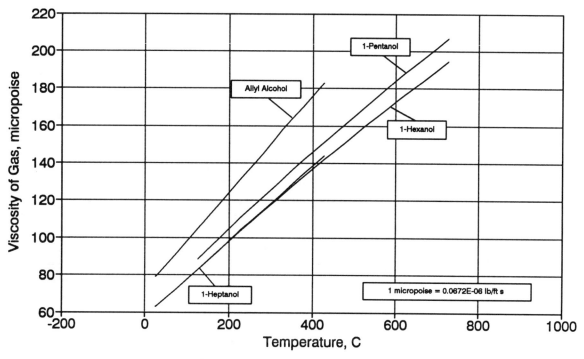

Figure 10-7 Viscosity of Gas

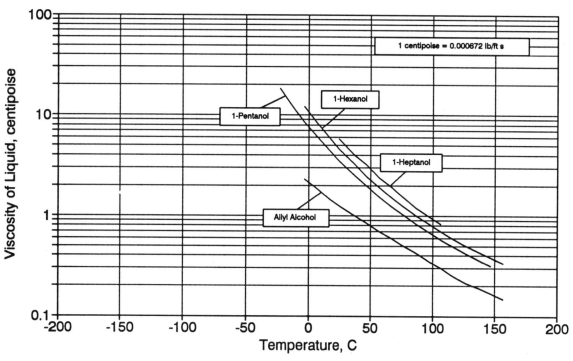

Figure 10-8 Viscosity of Liquid

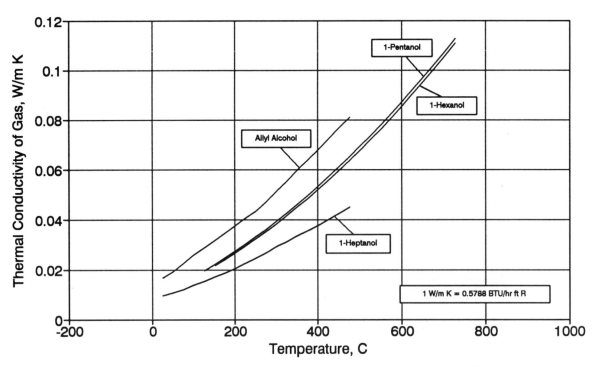

Figure 10-9 Thermal Conductivity of Gas

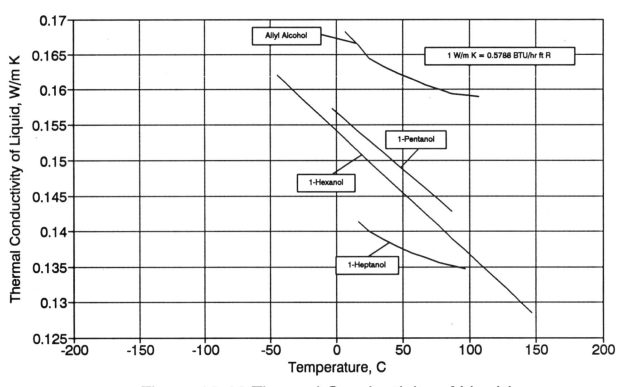

Figure 10-10 Thermal Conductivity of Liquid

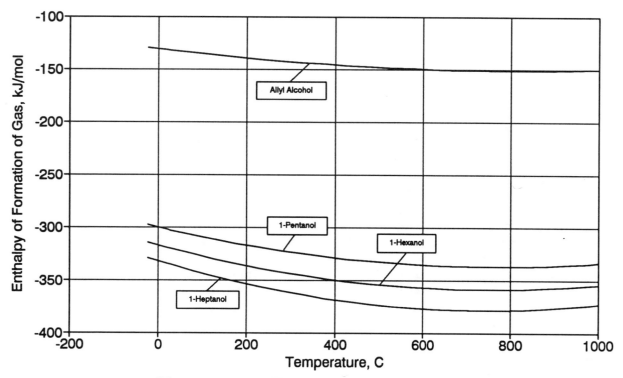

Figure 10-11 Enthalpy of Formation of Gas

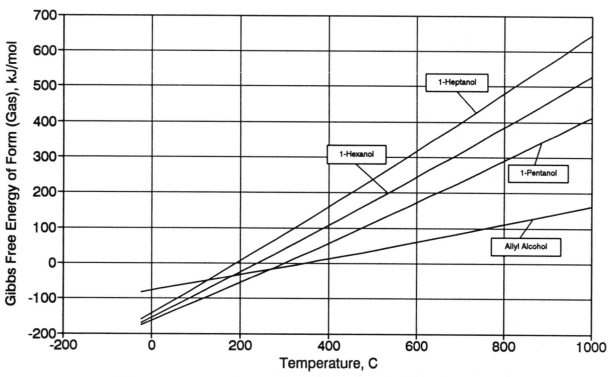

Figure 10-12 Gibbs Free Energy of Formation of Gas

Chapter 11

C$_2$ TO C$_4$ OXIDES

Robert W. Gallant, Carl L. Yaws and Xiang Pan

PHYSICAL PROPERTIES - Table 11-1

Property data from the literature (1-55,96,11,119,124,125,139,165,168,235-249) are given in Table 11-1. The critical constants were selected from the DIPPR project (5) except for butylene oxide (31). For ethylene and propylene oxides, the values are experimental. For the other compounds, the values are estimates. Additional property data such as acentric factor, enthalpy of formation, lower explosion limit in air and solubility in water are also available. The DIPPR (Design Institute for Physical Property Research) project (5) and recent data compilations by Yaws and co-workers (44-55) were consulted extensively in preparing the tabulation.

VAPOR PRESSURE - Figure 11-1

Results from the DIPPR project (5) were selected for vapor pressure from very low temperatures to the critical point except for butylene oxide (31). Correlation of data for vapor pressure as a function of temperature was accomplished using Equation (1-1). Results from this equation (Antoine-type with extended terms) are in favorable agreement with experimental data. Errors are about 1-10% or less in most cases.

HEAT OF VAPORIZATION - Figure 11-2

The data compilation of Yaws and co-workers (31,44,52) was selected for heat of vaporization for temperatures ranging from melting point to critical point except for epichlorohydrin (5). A modified Watson equation, Equation (1-2a), was used for correlation of the data as a function of temperature. Reliability of results is good with errors of about 1-10%.

LIQUID DENSITY - Figure 11-3

Results from the data compilation of Yaws and co-workers (31,44,54) were selected for liquid density from low temperatures at the melting point to higher temperatures up to the critical point except for epichlorohydrin (5). A modified Rackett equation, Equation (1-3a), was used for correlation of the data as a function of temperature. Results from the correlation are in favorable agreement with data. Deviations are less than 1-3% in most cases.

SURFACE TENSION - Figure 11-4

The data compilation Yaws and co-workers (31,44,55) was selected for surface tension for temperatures from melting point to critical point except for epichlorohydrin (5). Using data from the literature, correlation for surface tension as a function of temperature over the full liquid range was achieved by the modified Othmer equation, Equation (1-4a). Accuracy is good with errors being about 1-10% or less in most cases.

HEAT CAPACITY - Figures 11-5 and 11-6

Results from the data compilation of Yaws and co-workers (31,44) were selected for heat capacity of ideal gas except for epichlorohydrin (5). Correlation of data was accomplished using Equation (1-5a). Results are in favorable agreement with data. Errors are about 1-10% or less in most cases.

Results from the DIPPR project (5) were selected for heat capacity of liquid except for butylene oxide (31). The coverage applies to temperatures from below the boiling point to temperatures above the boiling point for most of the compounds. Data were correlated with a series expansion in temperature, Equation (1-6). Results are in fair agreement with data. Errors are about 10% or less using the correlation.

VISCOSITY - Figures 11-7 and 11-8

The DIPPR project (5) was selected for viscosity of gas except for butylene oxide (31). Data for gas viscosity as a function of temperature were correlated using Equation (1-7). Results are in fair agreement with data. Errors are about 10% or less in most cases.

Results from the data compilation of Yaws and co-workers (31) were selected for viscosity of liquid except for epichlorohydrin (5). Temperatures from below the boiling point to temperatures above the boiling point are covered for most of the compounds. Data for liquid viscosity as a function of temperature were correlated using the de Guzman - Andrade equation with extended terms, Equation (1-8). Correlation results and data are in agreement with errors being about 5-10% or less.

THERMAL CONDUCTIVITY - Figures 11-9 and 11-10

Results from the data compilation of Yaws and co-workers (31) were selected for thermal conductivity of gas except for epichlorohydrin (5). Data for gas thermal conductivity as a function of temperature were correlated using a series expansion in temperature. Reliability of results is rough with possible errors of 5-25%.

Results from the data compilation of Yaws and co-workers (31) were selected for thermal conductivity of liquid except for epichlorohydrin (30). The coverage applies to temperatures from below the boiling point to temperatures above the boiling point for most of the compounds. Data for liquid thermal conductivity as a function of temperature were correlated using a series expansion in temperature, Equation (1-10). Reliability of results is rough with possible errors of 5-100%.

ENTHALPY OF FORMATION - Figure 11-11

The data compilation of Yaws and co-workers (31,44,45) was selected for enthalpy of formation of ideal gas for all compounds except epichlorohydrin. For epichlorohydrin, the value at 25 C (5) was extended to higher temperatures by integration of the appropriate equations (177) which involve gas heat capacities. Data for enthalpy of formation of the ideal gas is a series expansion in temperature, Equation (1-11). Results from the correlation are in favorable agreement with data.

GIBB'S FREE ENERGY OF FORMATION - Figure 11-12

The data compilation of Yaws and co-workers (31,44,45) was selected for Gibb's free energy of formation of ideal gas for all compounds except epichlorohydrin. For epichlorohydrin, the value at 25 C (5) was extended to higher temperatures by integration of the appropriate equations (177) which involve gas heat capacities. Data for Gibb's free energy of formation of the ideal gas is a series expansion in temperature, Equation (1-12). Results from the correlation are in favorable agreement with data.

Table 11-1 Physical Properties

	Ethylene Oxide	Propylene Oxide	Butylene Oxide	Epichlorohydrin
1. Name				
2. Formula	C_2H_4O	C_3H_6O	C_4H_8O	C_3H_5ClO
3. Molecular Weight, g/mol	44.053	58.080	72.100	92.525
4. Critical Temperature, K	469.15	482.25	525.75	610.00
5. Critical Pressure, bar	71.941	49.244	47.100	49.000
6. Critical Volume, ml/mol	140.30	186.00	209.10	233.00
7. Critical Compressibility Factor	0.259	0.228	0.250	0.225
8. Acentric Factor	0.1979	0.2710	------	0.2562
9. Melting Point, K	161.45	161.22	123.15	215.95
10. Boiling Point @ 1 atm, K	283.85	307.05	336.35	389.26
11. Heat of Vaporization @ BP, kJ/kg	581.19	464.90	420.79	402.33
12. Density of Liquid @ 25 C, g/ml	0.8635	0.8219	0.8258	1.1761
13. Surface Tension @ 25 C, dynes/cm	23.23	21.23	23.15	36.29
14. Heat Capacity of Gas @ 25 C, J/g K	1.090	1.240	1.290	0.910
15. Heat Capacity of Liquid @ 25 C, J/g K	2.01	2.09	2.03	1.42
16. Viscosity of Gas @ 25 C, micropoise	93.04	87.73	78.56	------
17. Viscosity of Liquid @ 25 C, centipoise	0.249	0.301	0.396	1.073
18. Thermal Conductivity of Gas @ 25 C, W/m K	0.0131	0.0110	0.0081	------
19. Thermal Conductivity of Liquid @ 25 C, W/m K	0.153	0.150	0.152	0.073
20. Enthalpy of Formation of Gas @ 25 C, kJ/mol	-52.635	-92.759	-114.724	-107.800
21. Gibbs Free Energy of Formation of Gas @ 25 C, kJ/mol	-13.238	-25.773	-20.600	-36.744
22. Flash Point, K	------	235.93	------	304.26
23. Autoignition Temperature, K	702.04	738.15	------	689.15
24. Lower Explosion Limit in Air, vol %	3.0	2.1	------	3.8
25. Upper Explosion Limit in Air, vol %	------	21.5	------	21.0
26. Solubility in Water @ 25 C, ppm(wt)	total	259,300 @ 30 C	-----	------

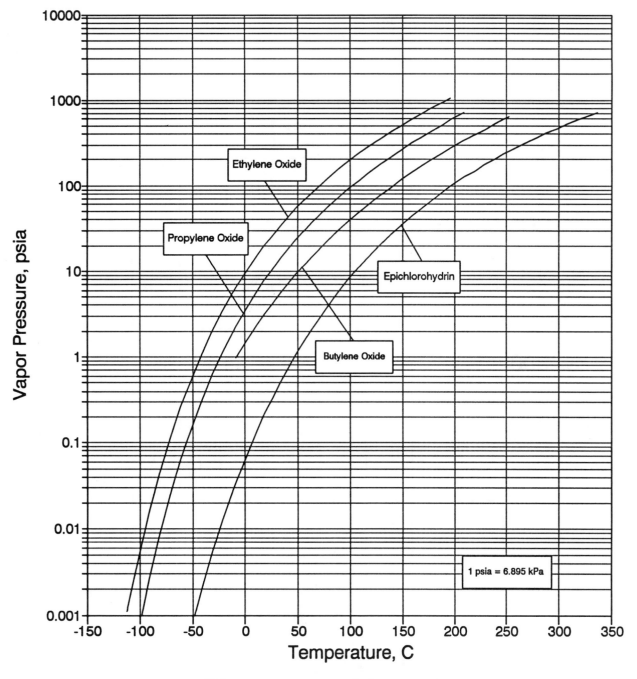

Figure 11-1 Vapor Pressure

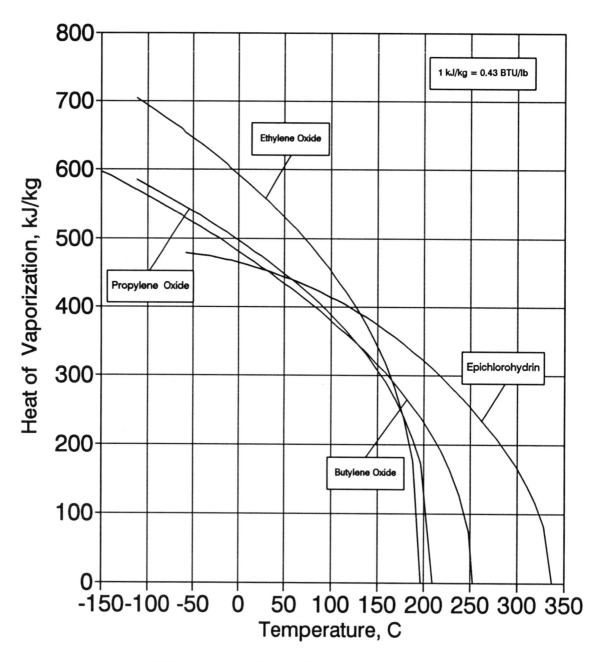

Figure 11-2 Heat of Vaporization

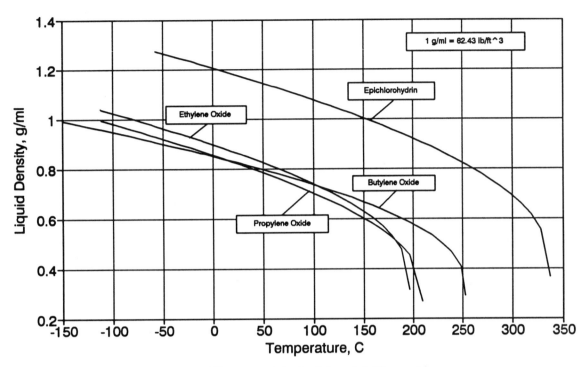

Figure 11-3 Liquid Density

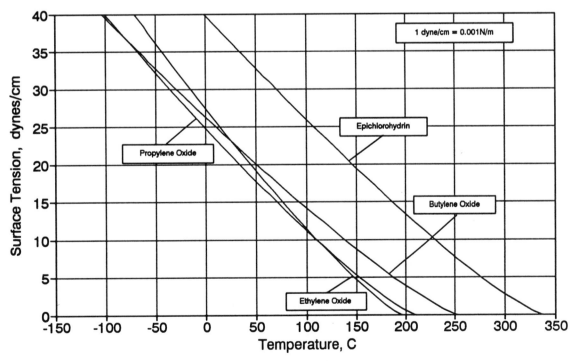

Figure 11-4 Surface Tension

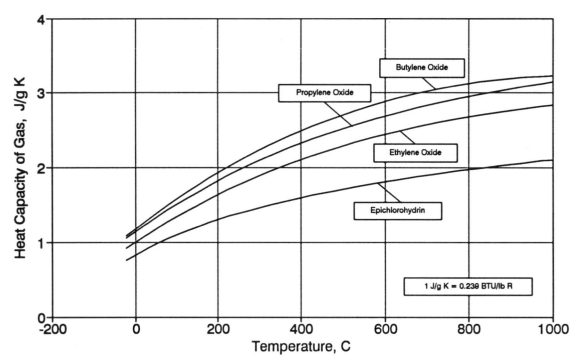

Figure 11-5 Heat Capacity of Gas

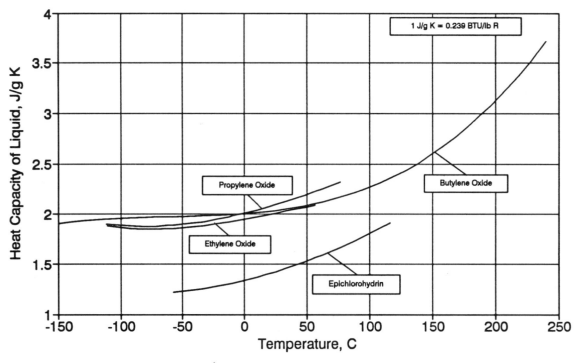

Figure 11-6 Heat Capacity of Liquid

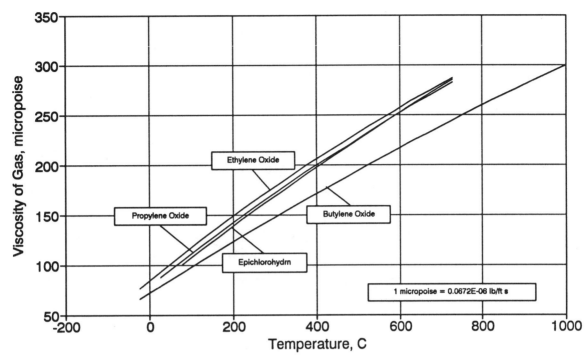

Figure 11-7 Viscosity of Gas

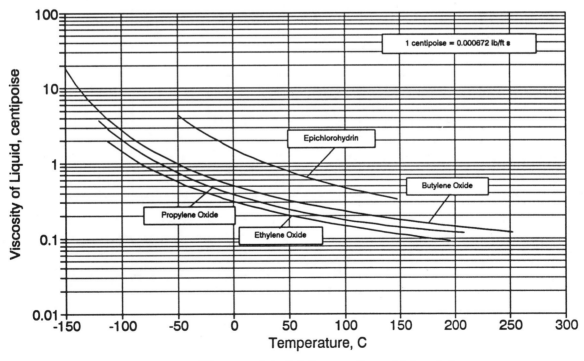

Figure 11-8 Viscosity of Liquid

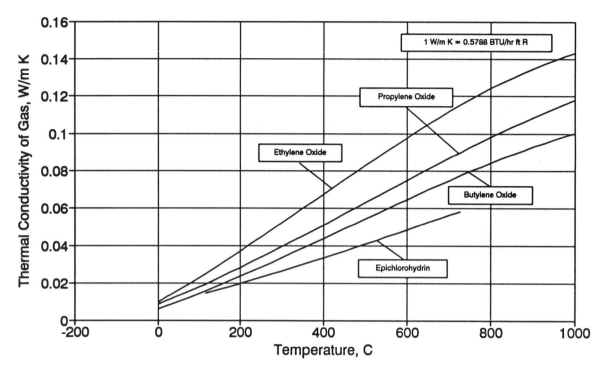

Figure 11-9 Thermal Conductivity of Gas

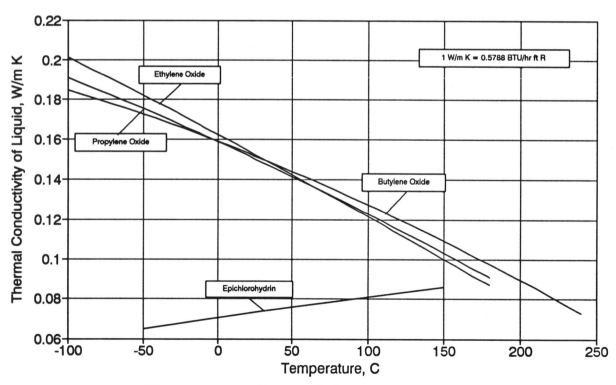

Figure 11-10 Thermal Conductivity of Liquid

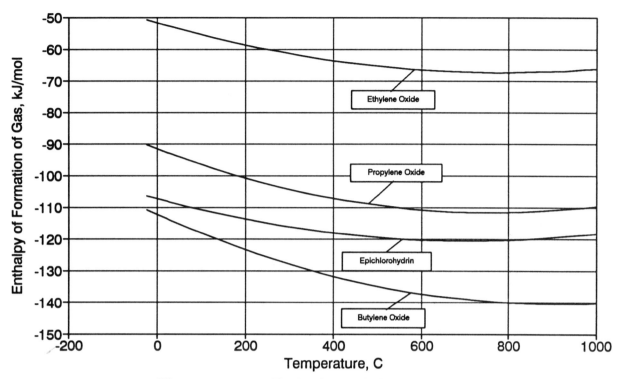

Figure 11-11 Enthalpy of Formation of Gas

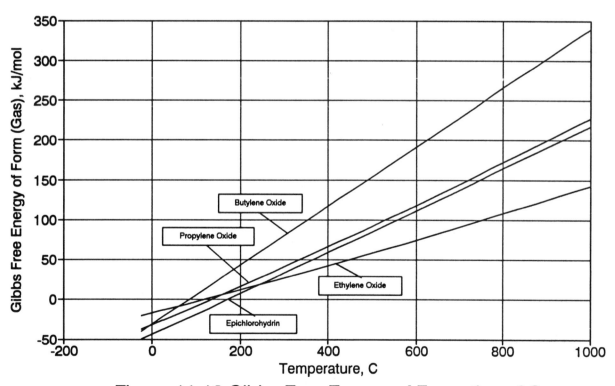

Figure 11-12 Gibbs Free Energy of Formation of Gas

Chapter 12

ETHYLENE GLYCOLS

Robert W. Gallant, Carl L. Yaws and Xiang Pan

PHYSICAL PROPERTIES - Table 12-1

Property data from the literature (1-55,110,111,113,114,125,137,162,165-177) are given in Table 12-1. The critical constants were selected from the DIPPR project (5). Additional property data such as acentric factor, enthalpy of formation, lower explosion limit in air and solubility in water are also available. The DIPPR (Design Institute for Physical Property Research) project (5) and recent data compilations by Yaws and co-workers (44-55) were consulted extensively in preparing the tabulation.

VAPOR PRESSURE - Figure 12-1

Results from the DIPPR project (5) were selected for vapor pressure from very low temperatures to the critical point. Correlation of data for vapor pressure as a function of temperature was accomplished using Equation (1-1). Results from this equation (Antoine-type with extended terms) are in favorable agreement with experimental data. Errors are about 1-10% or less in most cases.

HEAT OF VAPORIZATION - Figure 12-2

The data compilation of DIPPR project (5) was selected for heat of vaporization for temperatures ranging from melting point to critical point. A modified Watson equation, Equation (1-2a), was used for correlation of the data as a function of temperature. Reliability of results is fair with errors of about 1-10% or less except for triethylene glycol (possible 25%).

LIQUID DENSITY - Figure 12-3

Results from the data compilation of DIPPR project (5) were selected for liquid density from low temperatures at the melting point to higher temperatures up to the critical point. A modified Rackett equation, Equation (1-3a), was used for correlation of the data as a function of temperature. Results from the correlation are in favorable agreement with data. Deviations are less than 1-3% in most cases.

SURFACE TENSION - Figure 12-4

The data compilation of DIPPR project (5) was selected for surface tension for temperatures from melting point to critical point. Using data from the literature, correlation for surface tension as a function of temperature over the full liquid range was achieved by the modified Othmer equation, Equation (1-4a). Accuracy is good with errors being about 1-3% or less in most cases.

HEAT CAPACITY - Figures 12-5 and 12-6

Results from the data compilation of DIPPR project (5) were selected for heat capacity of ideal gas. Correlation of data was accomplished using Equation (1-5a). Results are in favorable agreement with data. Errors are about 1-10% or less in most cases.

Results from the DIPPR project (5) were selected for heat capacity of liquid. The coverage applies to temperatures from below the boiling point to temperatures above the boiling point for most of the compounds. Data were correlated with a series expansion in temperature, Equation (1-6). Results are in favorable agreement with data. Errors are about 3% or less using the correlation.

VISCOSITY - Figures 12-7 and 12-8

The DIPPR project (5) was selected for viscosity of gas. Data for gas viscosity as a function of temperature were correlated using Equation (1-7). Results are in fair agreement with data. Errors are about 10% or less in most cases.

The DIPPR project (5) was also selected for viscosity of liquid. Temperatures from below the boiling point to temperatures above the boiling point are covered for most of the compounds. Data for liquid viscosity as a function of temperature were correlated using the de Guzman - Andrade equation with extended terms, Equation (1-8). Correlation results and data are in fair agreement with errors being about 10% or less.

THERMAL CONDUCTIVITY - Figures 12-9 and 12-10

Results from the DIPPR project (5) were selected for thermal conductivity of gas. Data for gas thermal conductivity as a function of temperature were correlated using the Equation (1-9). Reliability of results is rough with possible errors of 25-50%.

Results from the DIPPR project (5) were selected for thermal conductivity of liquid. The coverage applies to temperatures from below the boiling point to temperatures above the boiling point for most of the compounds. Data for liquid thermal conductivity as a function of temperature were correlated using a series expansion in temperature, Equation (1-10). Results are in favorable agreement with data. Errors are about 1-3% or less in most cases.

ENTHALPY OF FORMATION - Figure 12-11

The data compilation of Yaws and co-workers (44,45) was selected for enthalpy of formation of ideal gas for ethylene glycol. For the other compounds, values at 25 C (5) were extended to higher temperatures by integration of the appropriate equations (177) which involve gas heat capacities. Data for enthalpy of formation of the ideal gas is a series expansion in temperature, Equation (1-11). Results from the correlation are in favorable agreement with data.

GIBB'S FREE ENERGY OF FORMATION - Figure 12-12

Results from the data compilation of Yaws and co-workers (44,46) were selected for Gibb's free energy of formation of ideal gas for ethylene glycol. For the other compounds, values at 25 C (5) were extended to higher temperatures by integration of the appropriate equations (177) which involve gas heat capacities. Data for Gibb's free energy of formation of the ideal gas is a series expansion in temperature, Equation (1-12). Correlation results are in favorable agreement with data.

Table 12-1 Physical Properties

	Ethylene Glycol	Diethylene Glycol	Triethylene Glycol
1. Name	Ethylene Glycol	Diethylene Glycol	Triethylene Glycol
2. Formula	C2H6O2	C4H10O3	C6H14O4
3. Molecular Weight, g/mol	62.068	106.121	150.174
4. Critical Temperature, K	645.00	680.00	700.00
5. Critical Pressure, bar	75.300	46.000	33.200
6. Critical Volume, ml/mol	191.00	312.00	443.00
7. Critical Compressibility Factor	0.268	0.254	0.253
8. Acentric Factor	1.1367	1.2006	1.3863
9. Melting Point, K	260.15	262.70	265.79
10. Boiling Point @ 1 atm, K	470.45	518.15	551.00
11. Heat of Vaporization @ BP, kJ/kg	845.76	580.10	427.76
12. Density of Liquid @ 25 C, g/ml	1.110	1.115	1.122
13. Surface Tension @ 25 C, dynes/cm	47.88	48.18	45.00
14. Heat Capacity of Gas @ 25 C, J/g K	1.56	1.26	1.27
15. Heat Capacity of Liquid @ 25 C, J/g K	2.39	2.31	2.19
16. Viscosity of Gas @ 25 C, micropoise	82.18	------	------
17. Viscosity of Liquid @ 25 C, centipoise	17.71	30.21	36.73
18. Thermal Conductivity of Gas @ 25 C, W/m K	------	------	0.0083
19. Thermal Conductivity of Liquid @ 25 C, W/m K	0.256	0.209	0.196
20. Enthalpy of Formation of Gas @ 25 C, kJ/mol	-389.32	-571.12	-725.09
21. Gibbs Free Energy of Formation of Gas @ 25 C, kJ/mol	-304.47	-409.08	-486.52
22. Flash Point, K	384.26	397.04	449.82
23. Autoignition Temperature, K	685.93	502.04	644.26
24. Lower Explosion Limit in Air, vol %	3.2	------	0.9
25. Upper Explosion Limit in Air, vol %	------	------	9.2
26. Solubility in Water @ 25 C, ppm(wt)	total	total	total

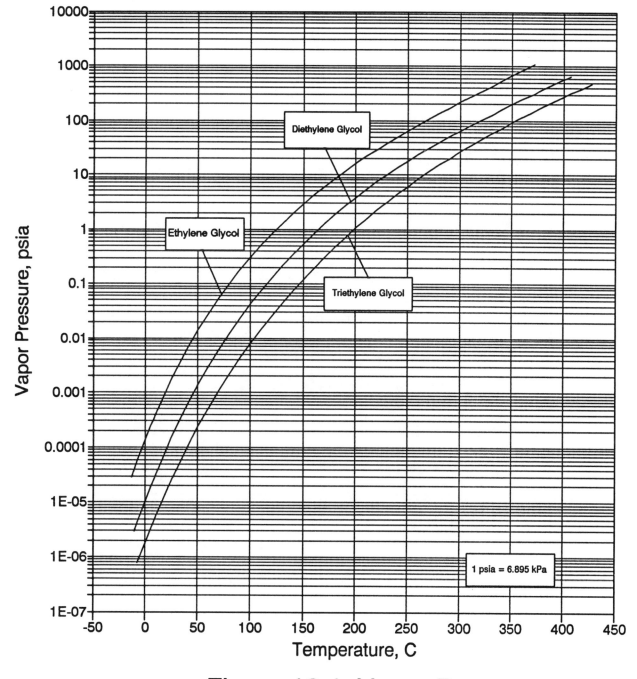

Figure 12-1 Vapor Pressure

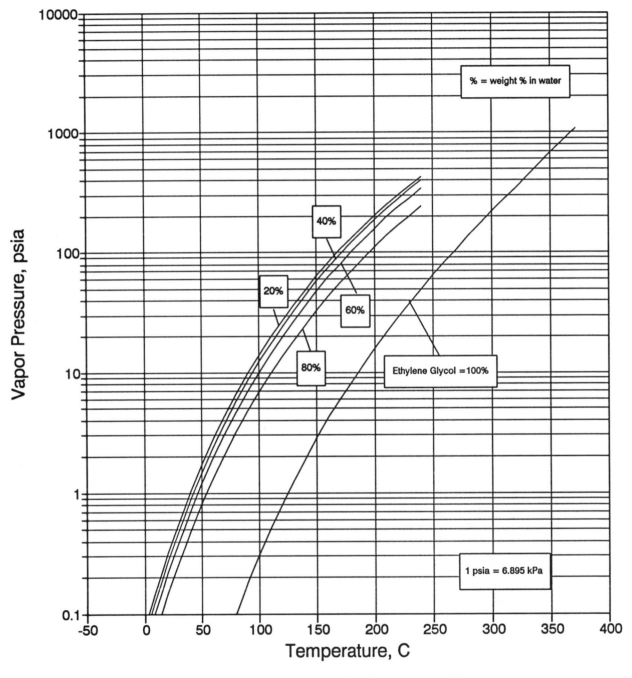

Figure 12-1A Vapor Pressure

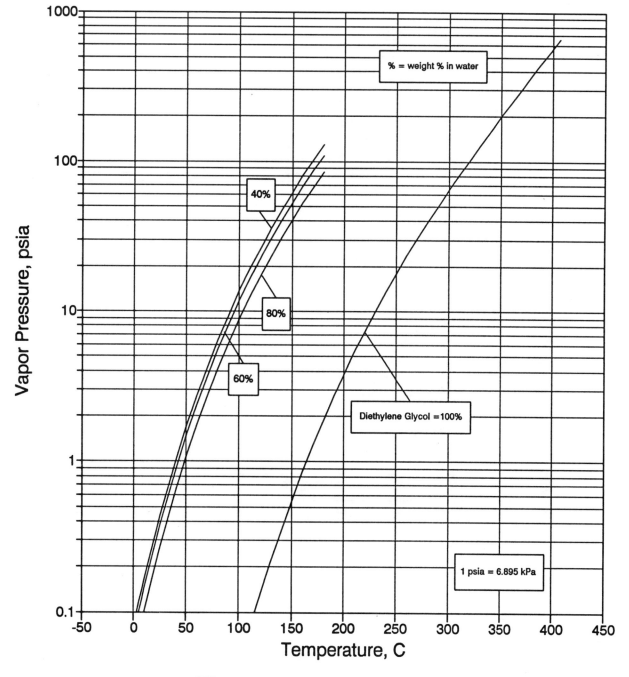

Figure 12-1B Vapor Pressure

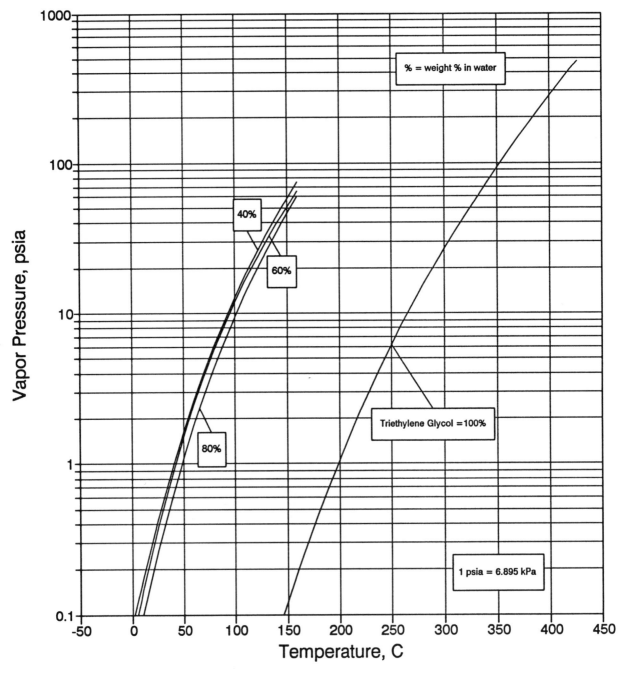

Figure 12-1C Vapor Pressure

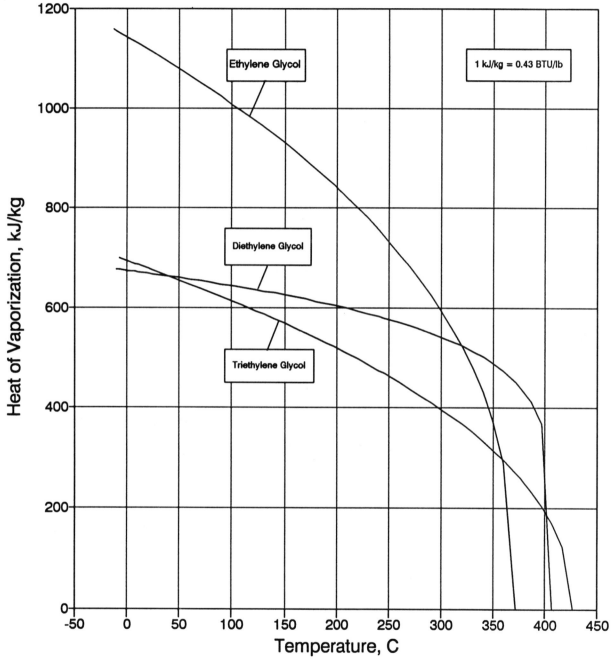

Figure 12-2 Heat of Vaporization

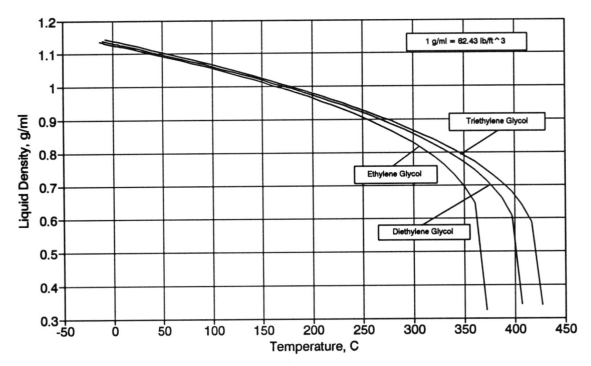

Figure 12-3 Liquid Density

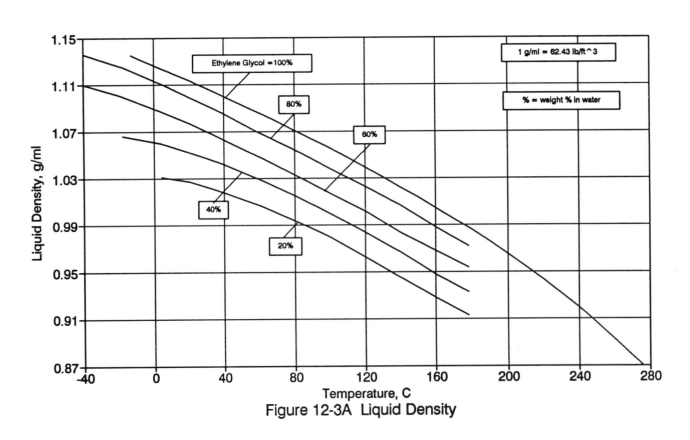

Figure 12-3A Liquid Density

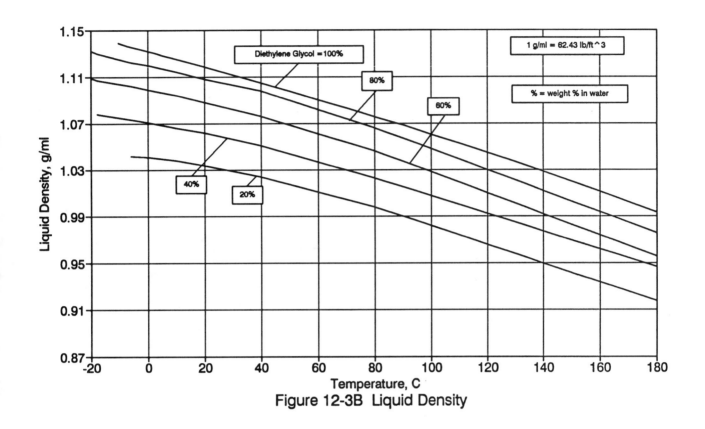
Figure 12-3B Liquid Density

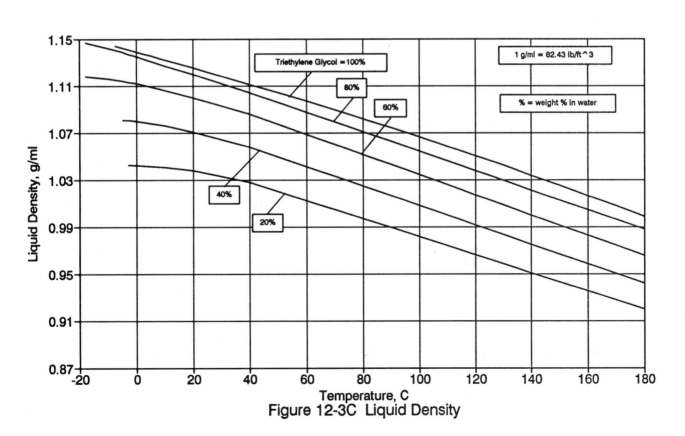
Figure 12-3C Liquid Density

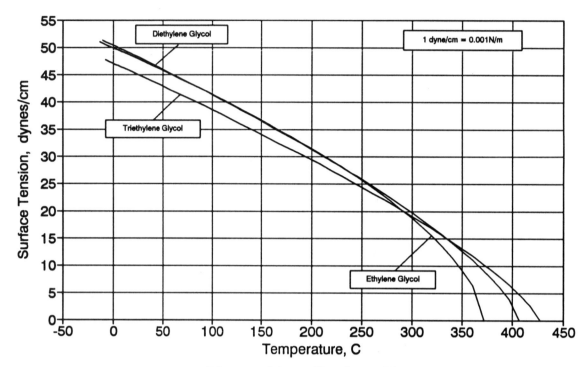

Figure 12-4 Surface Tension

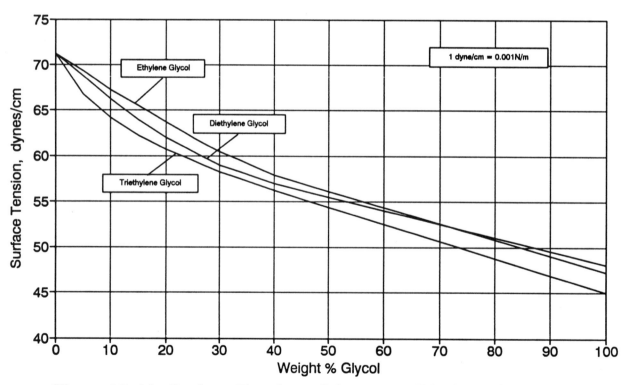

Figure 12-4A Surface Tension of Aqueous Ethylene Glycols at 25 C

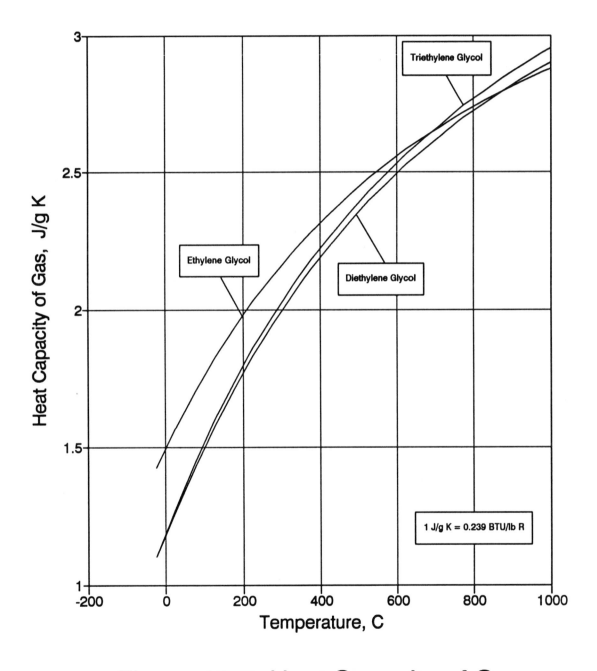

Figure 12-5 Heat Capacity of Gas

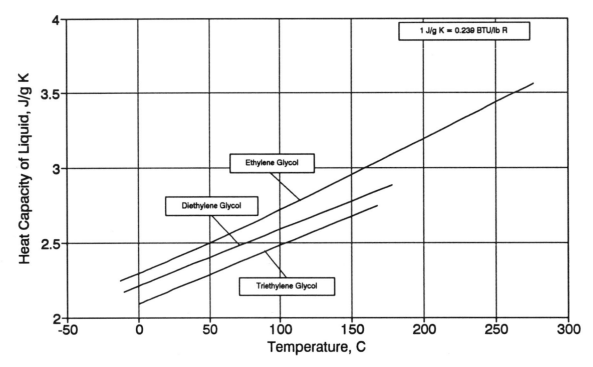

Figure 12-6 Heat Capacity of Liquid

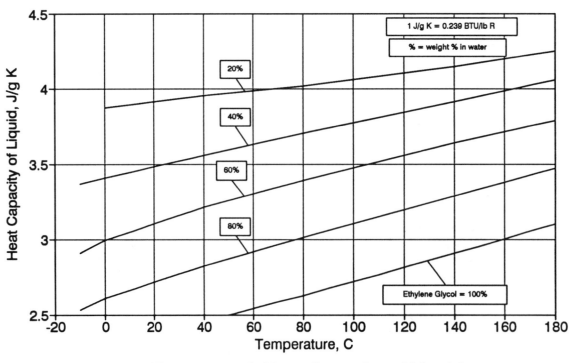

Figure 12-6A Heat Capacity of Liquid

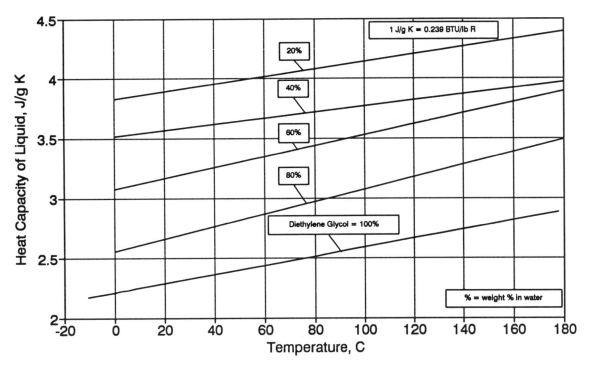

Figure 12-6B Heat Capacity of Liquid

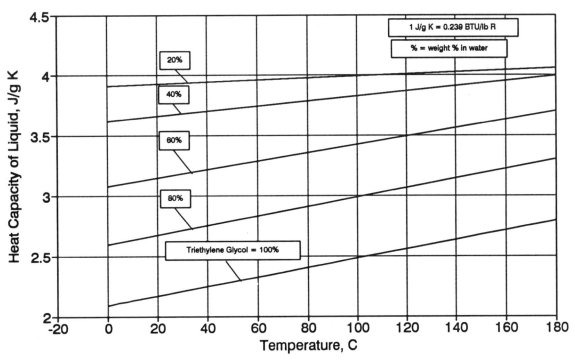

Figure 12-6C Heat Capacity of Liquid

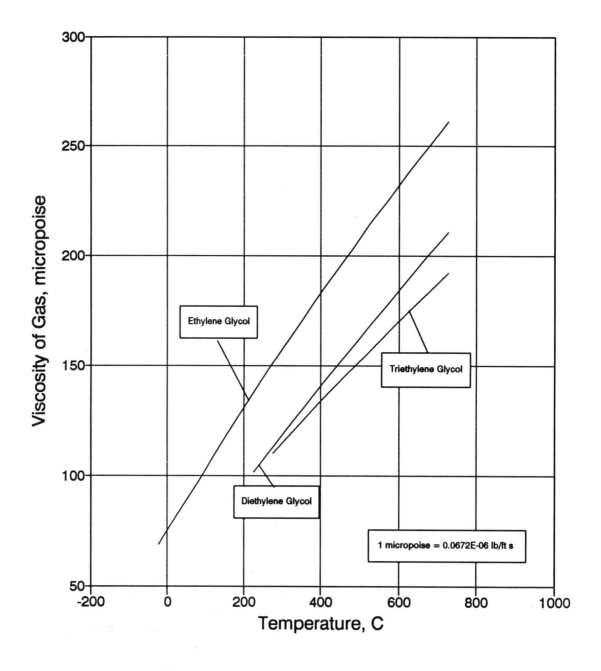

Figure 12-7 Viscosity of Gas

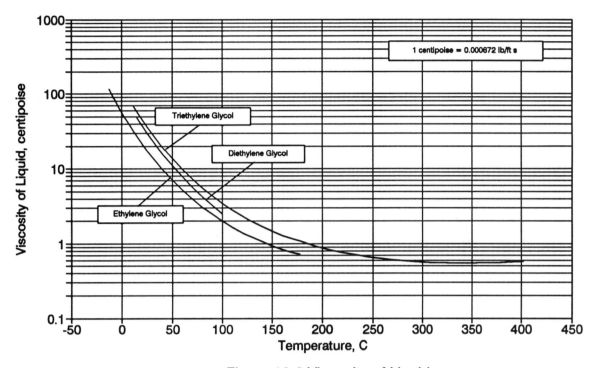

Figure 12-8 Viscosity of Liquid

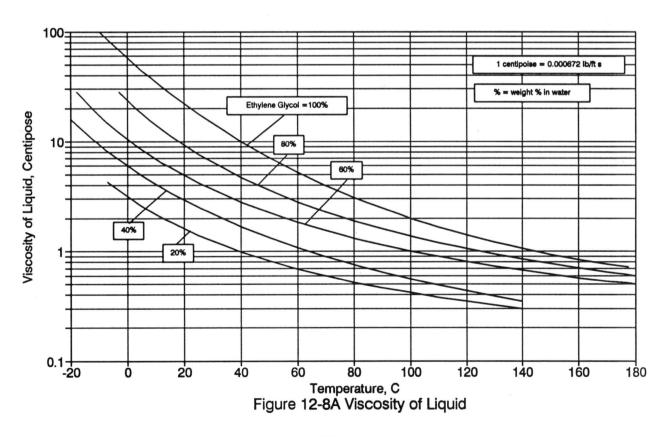

Figure 12-8A Viscosity of Liquid

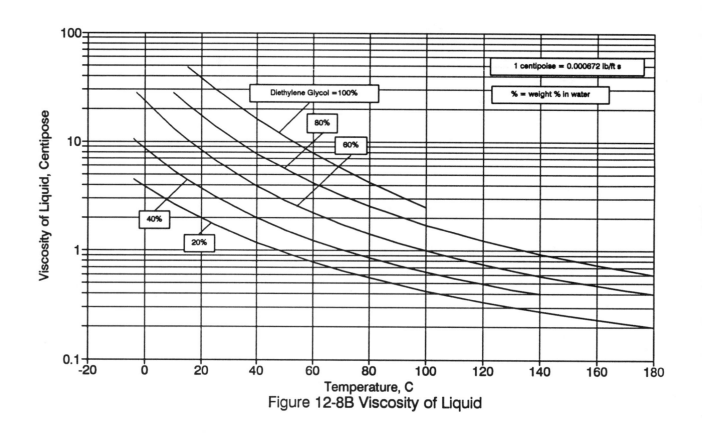
Figure 12-8B Viscosity of Liquid

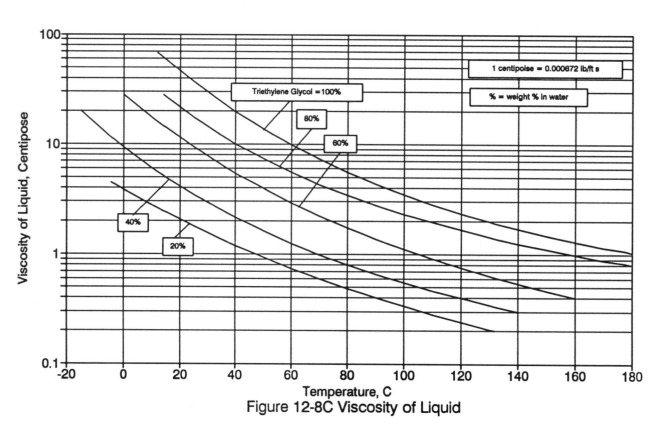
Figure 12-8C Viscosity of Liquid

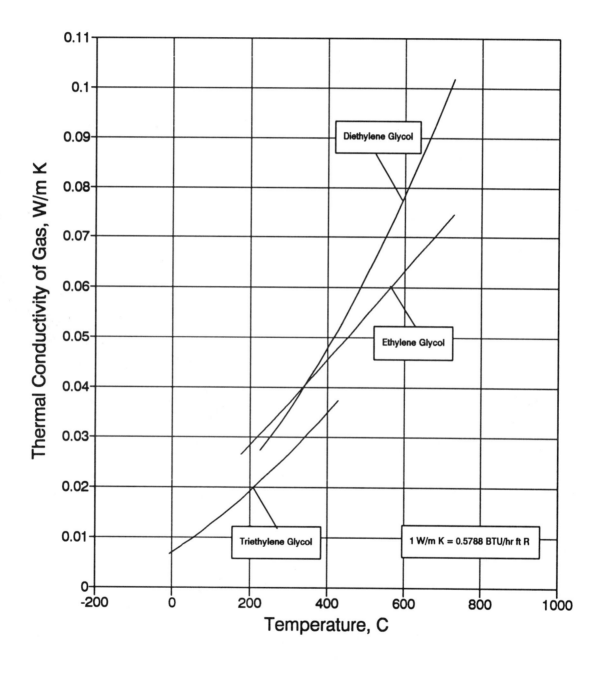

Figure 12-9 Thermal Conductivity of Gas

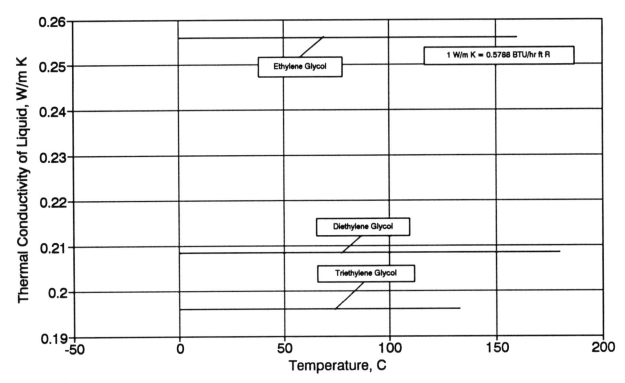

Figure 12-10 Thermal Conductivity of Liquid

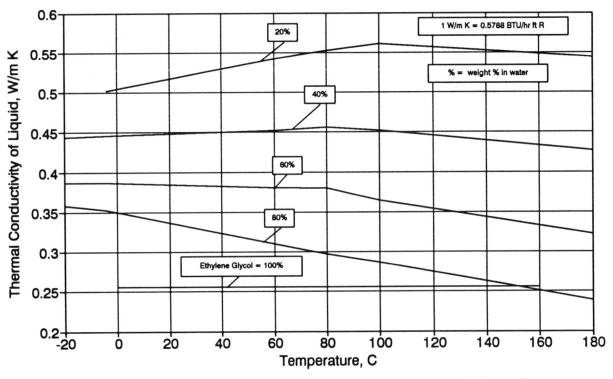

Figure 12-10A Thermal Conductivity of Liquid

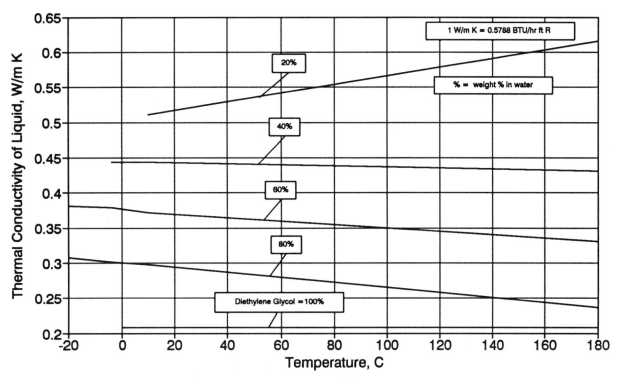

Figure 12-10B Thermal Conductivity of Liquid

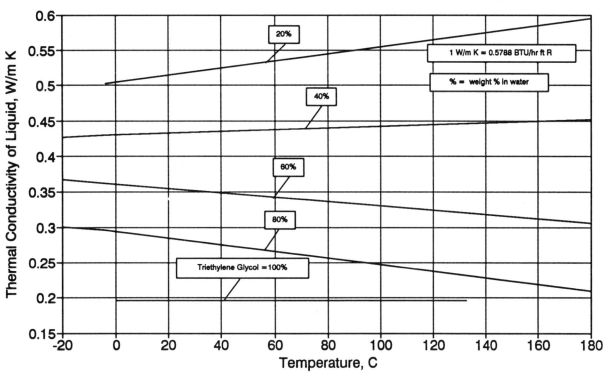

Figure 12-10C Thermal Conductivity of Liquid

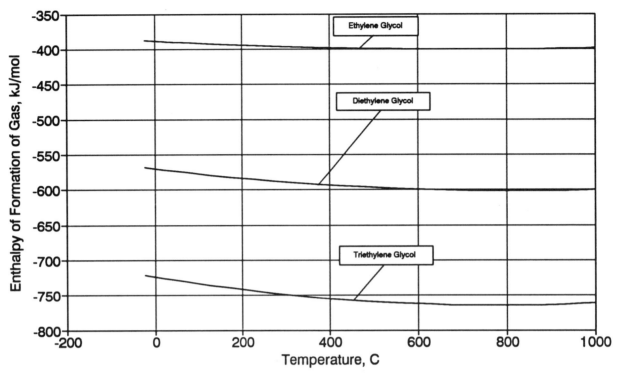

Figure 12-11 Enthalpy of Formation of Gas

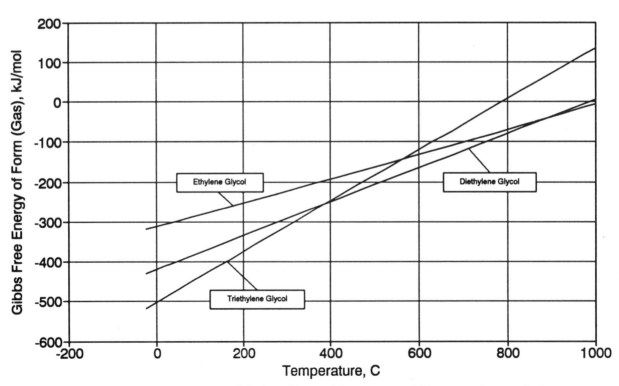

Figure 12-12 Gibbs Free Energy of Formation of Gas

Chapter 13

PROPYLENE GLYCOLS AND GLYCERINE

Robert W. Gallant, Carl L. Yaws and Xiang Pan

PHYSICAL PROPERTIES - Table 13-1

Property data from the literature (1-55,111,117,118,137,162,165,172,175,200-218) are given in Table 13-1. The critical constants were selected from the DIPPR project (5) except for dipropylene glycol (30). Additional property data such as acentric factor, enthalpy of formation, lower explosion limit in air and solubility in water are also available. The DIPPR (Design Institute for Physical Property Research) project (5) and recent data compilations by Yaws and co-workers (44-55) were consulted extensively in preparing the tabulation.

VAPOR PRESSURE - Figure 13-1

Results from the DIPPR project (5) were selected for vapor pressure from very low temperatures to the critical point except for dipropylene glycol (30). Correlation of data for vapor pressure as a function of temperature was accomplished using Equation (1-1). Results from this equation (Antoine-type with extended terms) are in favorable agreement with experimental data. Errors are about 1-10% or less in most cases.

HEAT OF VAPORIZATION - Figure 13-2

The data compilation of DIPPR project (5) was selected for heat of vaporization for temperatures ranging from melting point to critical point except for dipropylene glycol (30). A modified Watson equation, Equation (1-2a), was used for correlation of the data as a function of temperature. Reliability of results is good with errors of about 1-10%.

LIQUID DENSITY - Figure 13-3

Results from the data compilation of DIPPR project (5) were selected for liquid density from low temperatures at the melting point to higher temperatures up to the critical point except for dipropylene glycol (30). A modified Rackett equation, Equation (1-3a), was used for correlation of the data as a function of temperature. Results from the correlation are in favorable agreement with data. Deviations are less than 1-3% in most cases.

SURFACE TENSION - Figure 13-4

The data compilation of DIPPR project (5) was selected for surface tension for temperatures from melting point to critical point except for dipropylene glycol (30). Using data from the literature, correlation for surface tension as a function of temperature over the full liquid range was achieved by the modified Othmer equation, Equation (1-4a). Accuracy is good with errors being about 1-10% or less in most cases.

HEAT CAPACITY - Figures 13-5 and 13-6

Results from the data compilation of DIPPR project (5) were selected for heat capacity of ideal gas except for dipropylene glycol (30). Correlation of data was accomplished using Equation (1-5a). Results are in favorable agreement with data. Errors are about 1-10% or less in most cases.

Results from the DIPPR project (5) were selected for heat capacity of liquid except for dipropylene glycol (30). The coverage applies to temperatures from below the boiling point to temperatures above the boiling point for most of the compounds. Data were correlated with a series expansion in temperature, Equation (1-6). Results are in favorable agreement with data. Errors are about 3% or less using the correlation.

VISCOSITY - Figures 13-7 and 13-8

The DIPPR project (5) was selected for viscosity of gas except for dipropylene glycol (30). Data for gas viscosity as a function of temperature were correlated using Equation (1-7). Results are in fair agreement with data. Errors are about 10% or less in most cases.

The DIPPR project (5) was also selected for viscosity of liquid except for dipropylene glycol (30). Temperatures from below the boiling point to temperatures above the boiling point are covered for most of the compounds. Data for liquid viscosity as a function of temperature were correlated using the de Guzman - Andrade equation with extended terms, Equation (1-8). Correlation results and data are in agreement with errors being about 3-10% or less.

THERMAL CONDUCTIVITY - Figures 13-9 and 13-10

Results from the DIPPR project (5) were selected for thermal conductivity of gas except for dipropylene glycol (30). Data for gas thermal conductivity as a function of temperature were correlated using the Equation (1-9). Reliability of results is rough with possible errors of 25%.

Results from the DIPPR project (5) were selected for thermal conductivity of liquid except for dipropylene glycol (30). The coverage applies to temperatures from below the boiling point to temperatures above the boiling point for most of the compounds. Data for liquid thermal conductivity as a function of temperature were correlated using a series expansion in temperature, Equation (1-10). Results are in favorable agreement with data. Errors are about 1-5% or less in most cases.

ENTHALPY OF FORMATION - Figure 13-11

For propylene glycol and glycerine, values at 25 C are available (5). For dipropylene glycol, the value at 25 C was estimated using the group contribution of Joback (25). Values at 25 C (5) were extended to higher temperatures by integration of the appropriate equations (177) which involve gas heat capacities. Data for enthalpy of formation of the ideal gas is a series expansion in temperature, Equation (1-11). Results from the correlation are in favorable agreement with data.

GIBB'S FREE ENERGY OF FORMATION - Figure 13-12

For propylene glycol and glycerine, values at 25 C are available (5). For dipropylene glycol, the value at 25 C was estimated using the group contribution of Joback (25). Values at 25 C (5) were extended to higher temperatures by integration of the appropriate equations (177) which involve gas heat capacities. Data for Gibb's free energy of formation of the ideal gas is a series expansion in temperature, Equation (1-12). Correlation results are in favorable agreement with data.

Table 13-1 Physical Properties

	Propylene Glycol	Dipropylene Glycol	Glycerine
1. Name	Propylene Glycol	Dipropylene Glycol	Glycerine
2. Formula	$C_3H_8O_2$	$C_6H_{14}O_3$	$C_3H_8O_3$
3. Molecular Weight, g/mol	76.095	134.180	92.094
4. Critical Temperature, K	626.00	651.15	723.00
5. Critical Pressure, bar	61.000	35.850	40.000
6. Critical Volume, ml/mol	239.00	418.00	264.00
7. Critical Compressibility Factor	0.280	0.276	0.176
8. Acentric Factor	1.1065	------	1.3196
9. Melting Point, K	213.15	------	291.33
10. Boiling Point @ 1 atm, K	460.75	504.15	563.15
11. Heat of Vaporization @ BP, kJ/kg	715.95	405.98	718.05
12. Density of Liquid @ 25 C, g/ml	1.033	1.021	1.257
13. Surface Tension @ 25 C, dynes/cm	35.47	32.83	63.43
14. Heat Capacity of Gas @ 25 C, J/g K	1.34	1.32	1.30
15. Heat Capacity of Liquid @ 25 C, J/g K	2.51	2.44	2.38
16. Viscosity of Gas @ 25 C, micropoise	-----	67.60	------
17. Viscosity of Liquid @ 25 C, centipoise	40.39	86.65	923.50
18. Thermal Conductivity of Gas @ 25 C, W/m K	------	------	------
19. Thermal Conductivity of Liquid @ 25 C, W/m K	0.200	0.172	0.292
20. Enthalpy of Formation of Gas @ 25 C, kJ/mol	-421.50	-616.41	-582.80
21. Gibbs Free Energy of Formation of Gas @ 25 C, kJ/mol	-304.48	-383.88	-448.49
22. Flash Point, K	372.04	------	433.15
23. Autoignition Temperature, K	684.26	------	665.93
24. Lower Explosion Limit in Air, vol %	2.6	------	------
25. Upper Explosion Limit in Air, vol %	12.5	------	------
26. Solubility in Water @ 25 C, ppm(wt)	total	total	total

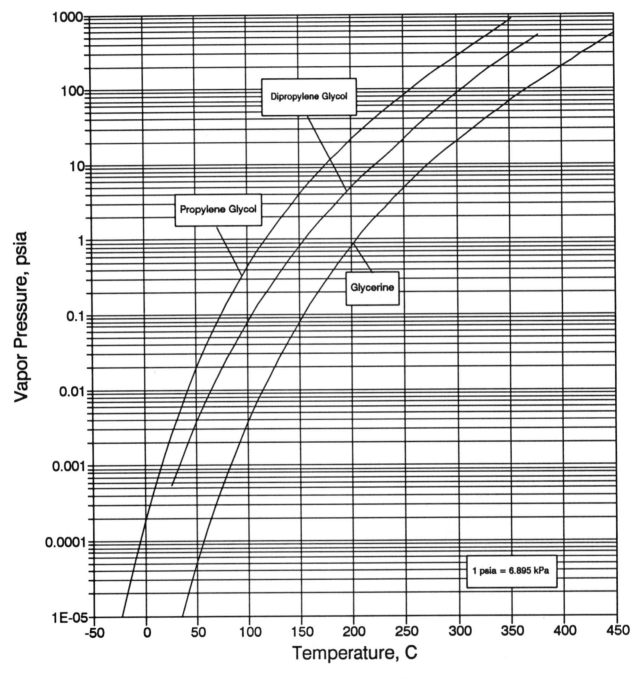

Figure 13-1 Vapor Pressure

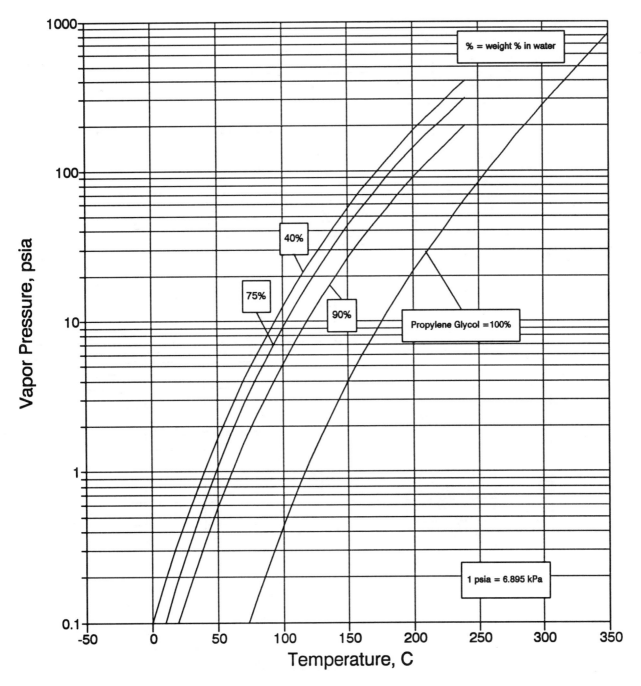

Figure 13-1A Vapor Pressure

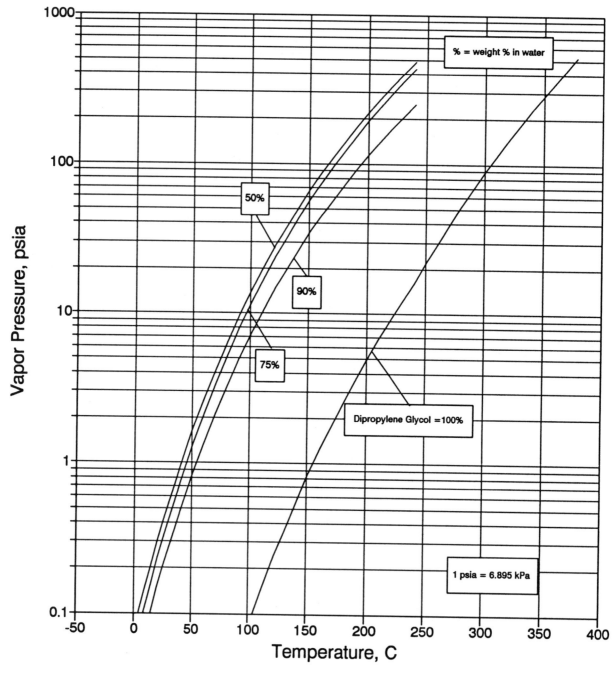

Figure 13-1B Vapor Pressure

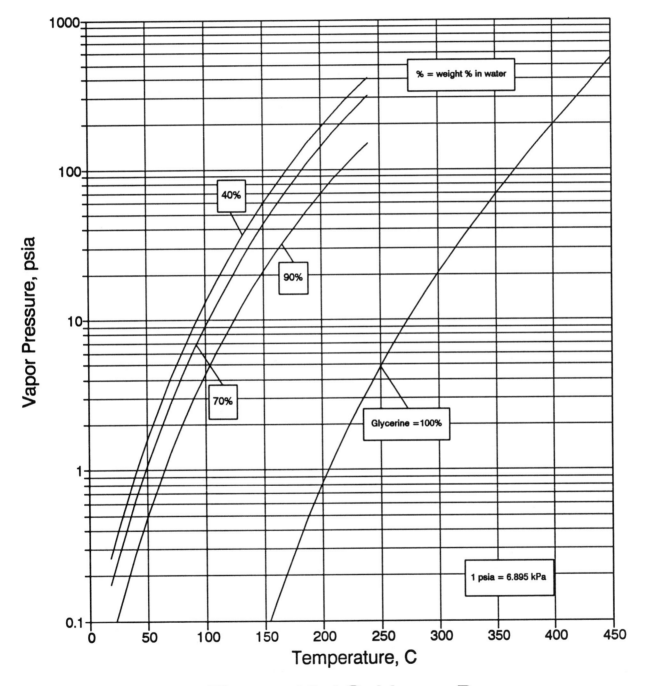

Figure 13-1C Vapor Pressure

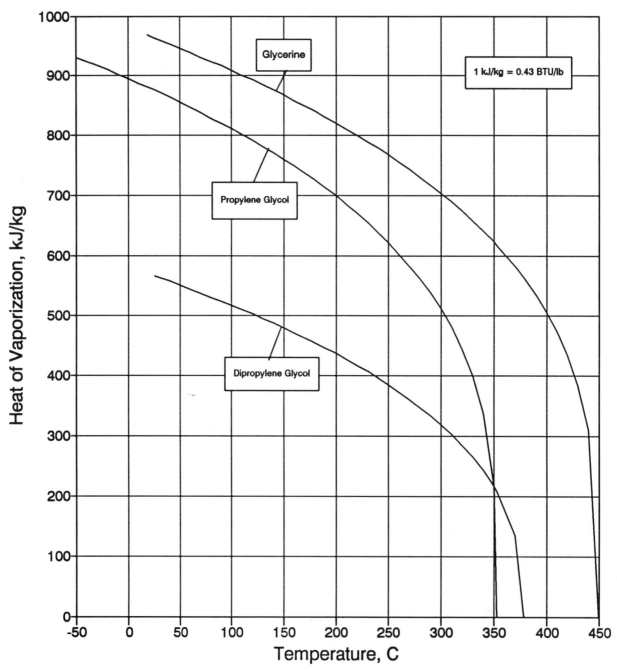

Figure 13-2 Heat of Vaporization

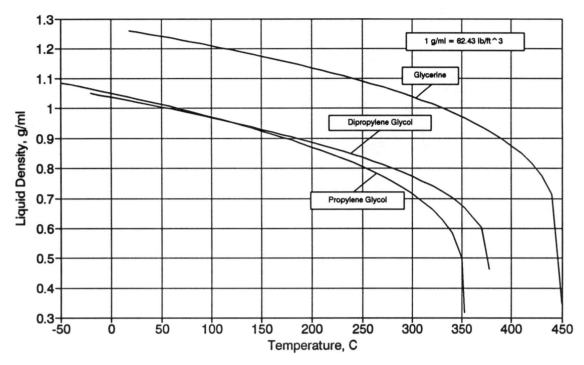

Figure 13-3 Liquid Density

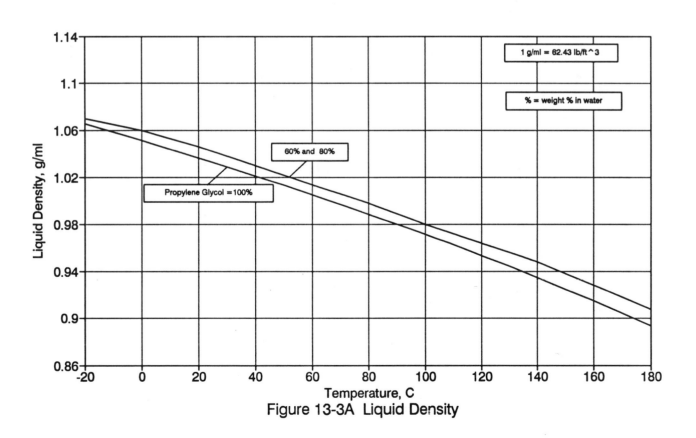

Figure 13-3A Liquid Density

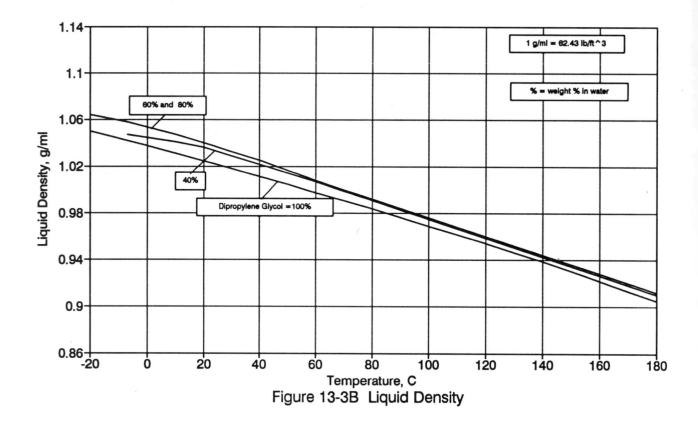
Figure 13-3B Liquid Density

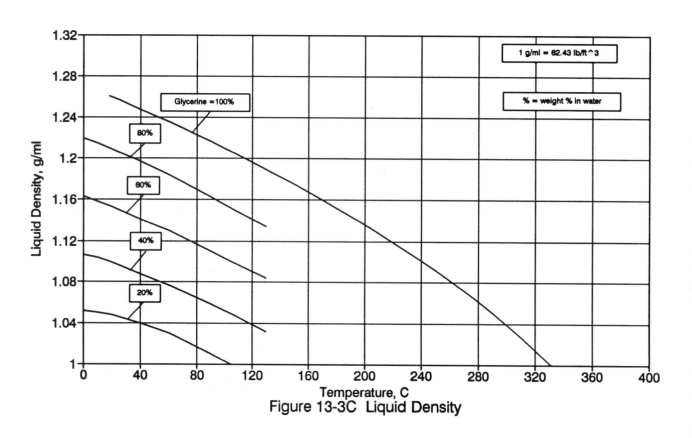
Figure 13-3C Liquid Density

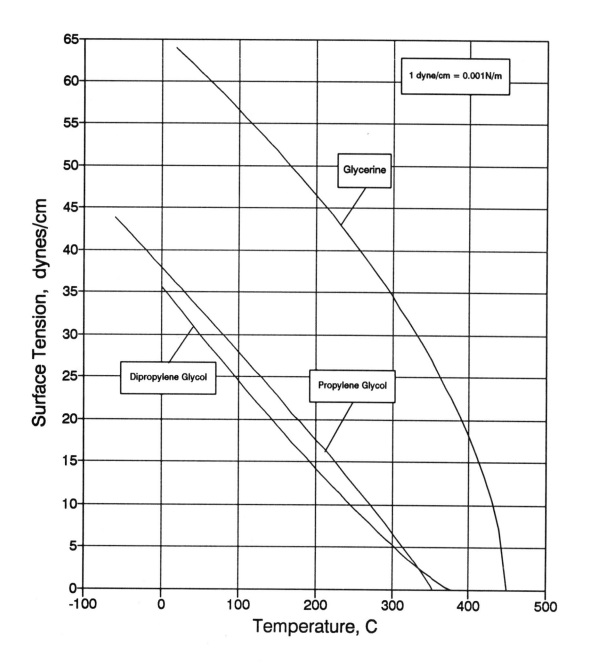

Figure 13-4 Surface Tension

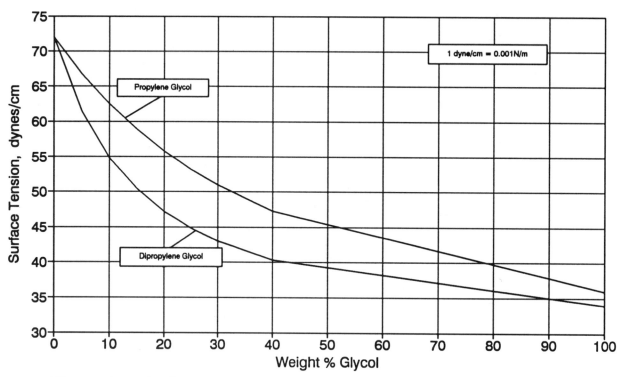

Figure 13-4A Surface Tension of Aqueous Ethylene Glycols at 25 C

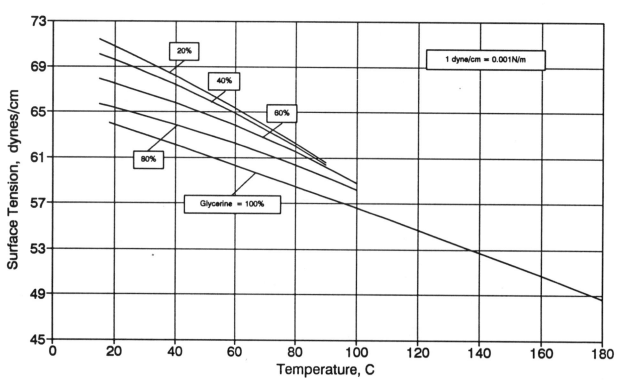

Figure 13-4B Surface Tension of Aqueous Ethylene Glycols at 25 C

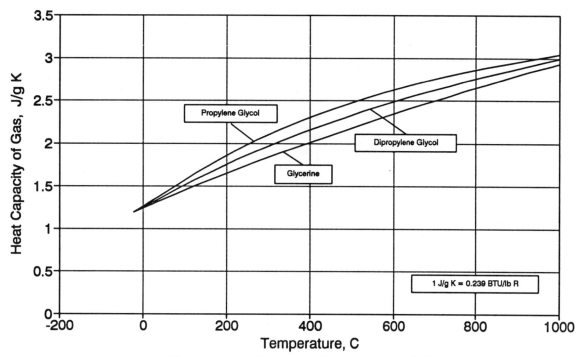

Figure 13-5 Heat Capacity of Gas

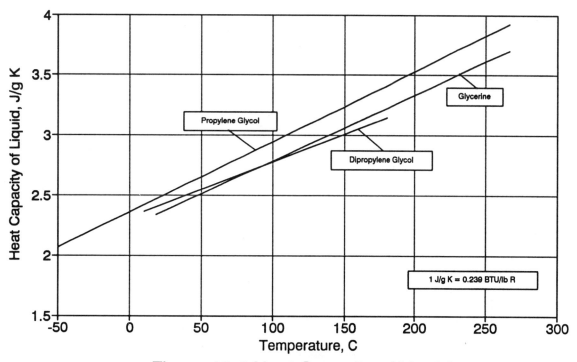

Figure 13-6 Heat Capacity of Liquid

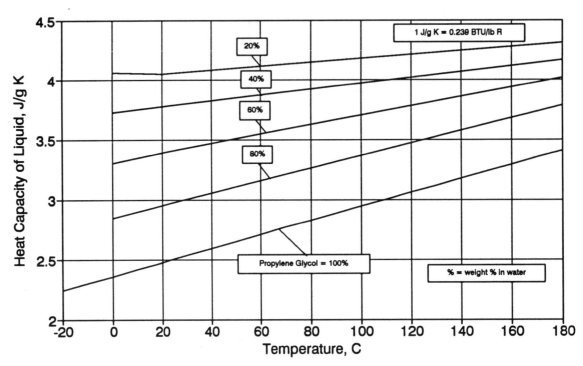

Figure 13-6A Heat Capacity of Liquid

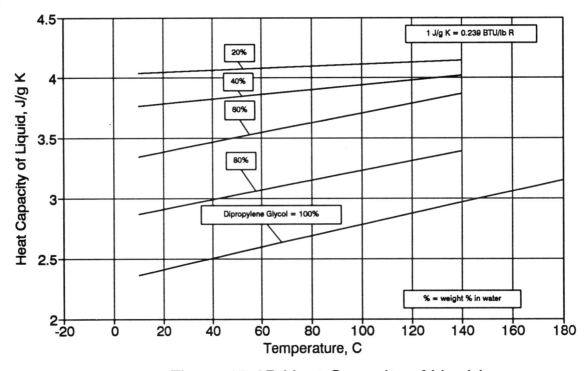

Figure 13-6B Heat Capacity of Liquid

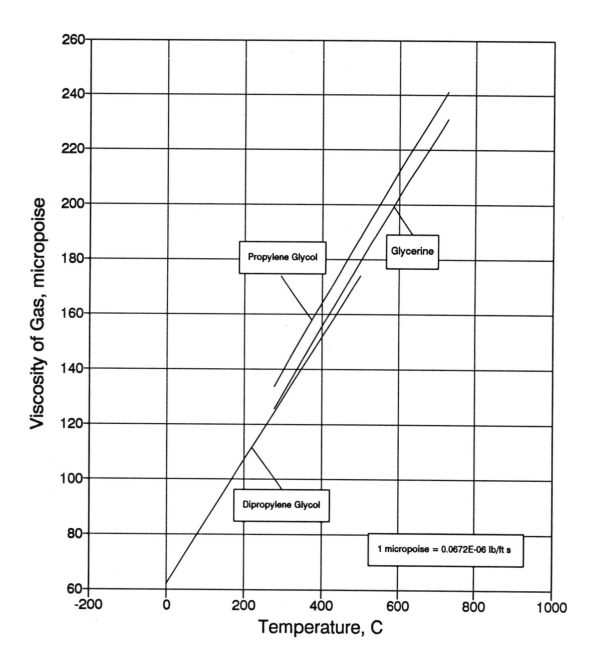

Figure 13-7 Viscosity of Gas

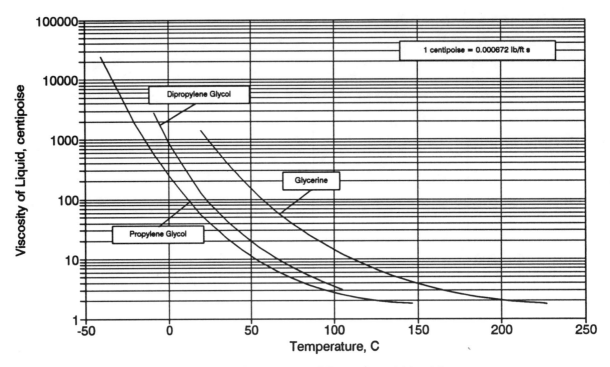

Figure 13-8 Viscosity of Liquid

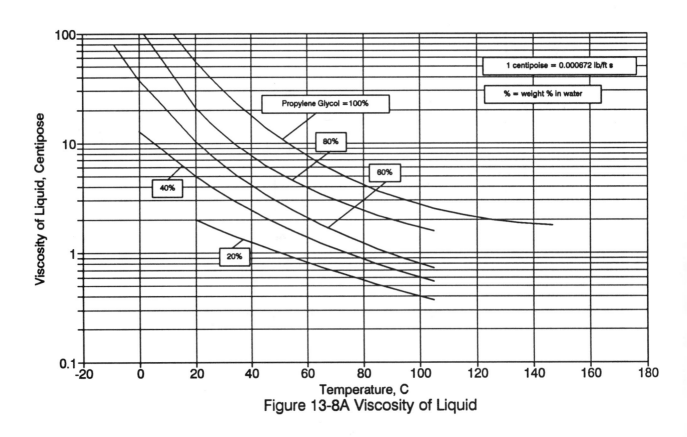

Figure 13-8A Viscosity of Liquid

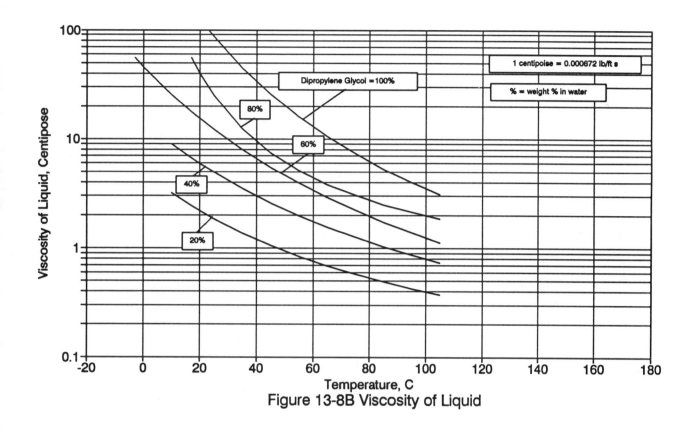
Figure 13-8B Viscosity of Liquid

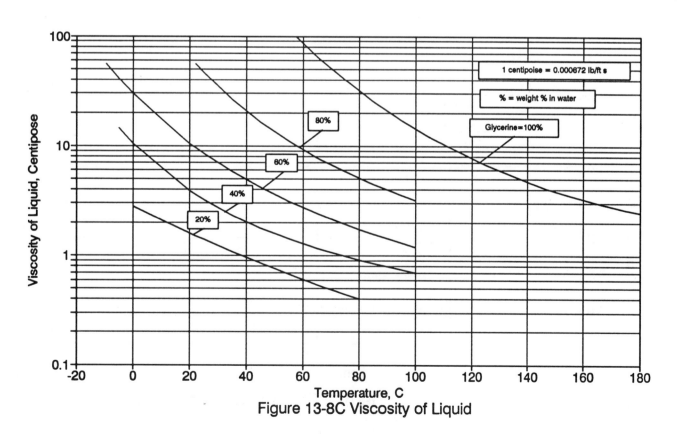
Figure 13-8C Viscosity of Liquid

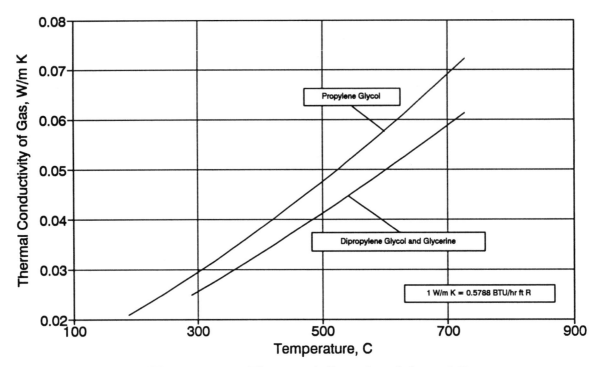

Figure 13-9 Thermal Conductivity of Gas

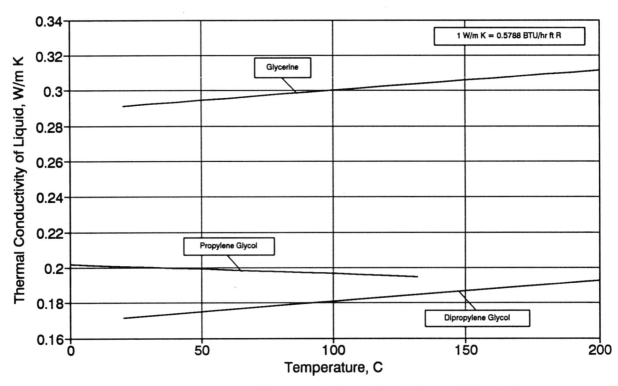

Figure 13-10 Thermal Conductivity of Liquid

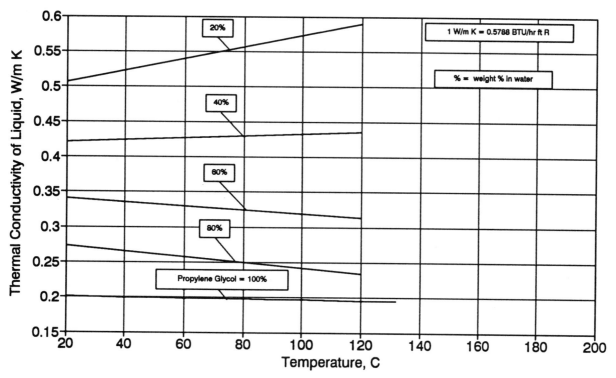

Figure 13-10A Thermal Conductivity of Liquid

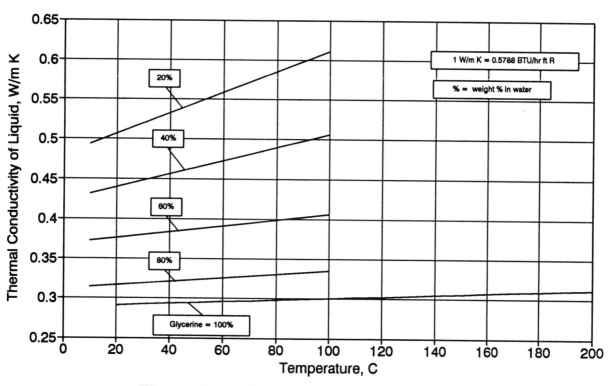

Figure 13-10B Thermal Conductivity of Liquid

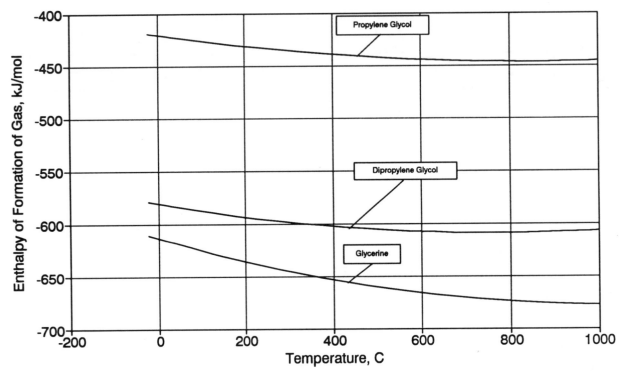

Figure 13-11 Enthalpy of Formation of Gas

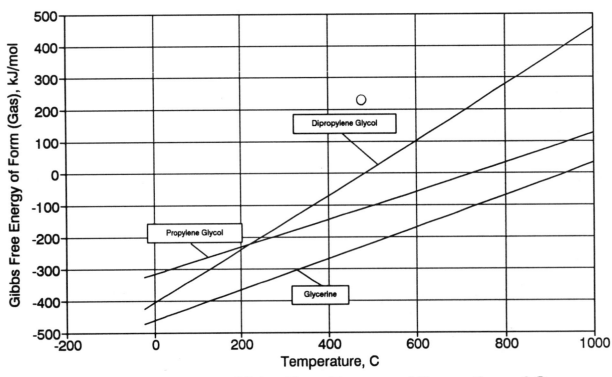

Figure 13-12 Gibbs Free Energy of Formation of Gas

Chapter 14

C_5 TO C_8 ALKANES

Robert W. Gallant and Carl L. Yaws

PHYSICAL PROPERTIES - Table 14-1

Property data from the literature (1-55,57,58,62,73,118,124,125,162,178-199) are given in Table 14-1. The critical constants have been determined experimentally and are available (1-7). Additional property data such as acentric factor, enthalpy of formation, lower explosion limit in air and solubility in water are also available. The DIPPR (Design Institute for Physical Property Research) project (5) and recent data compilations by Yaws and co-workers (44-55) were consulted extensively in preparing the tabulation.

VAPOR PRESSURE - Figure 14-1

Results from the DIPPR project (5) were selected for vapor pressure from very low temperatures to the critical point. Correlation of data for vapor pressure as a function of temperature was accomplished using Equation (1-1). Results from this equation (Antoine-type with extended terms) are in favorable agreement with experimental data. Errors are about 1-3% or less in most cases.

HEAT OF VAPORIZATION - Figure 14-2

The data compilation of Yaws and co-workers (44,52) was selected for heat of vaporization for temperatures ranging from melting point to critical point. The Watson equation, Equation (1-2), was used for correlation of the data as a function of temperature. Reliability of results is good with errors of about 1-5% or less.

LIQUID DENSITY - Figure 14-3

Results from the data compilation of Yaws and co-workers (44,54) were selected for liquid density from low temperatures at the melting point to higher temperatures up to the critical point. A modified Rackett equation, Equation (1-3), was used for correlation of the data as a function of temperature. Results from the correlation are in favorable agreement with data. Deviations are less than 1-2% in most cases.

SURFACE TENSION - Figure 14-4

The data compilation of Yaws and co-workers (44,55) was selected for surface tension for temperatures from melting point to critical point. Using data from the literature, correlation for surface tension as a function of temperature over the full liquid range was achieved by the modified Othmer equation, Equation (1-4). Accuracy is good with errors being about 1-3% or less in most cases.

HEAT CAPACITY - Figures 14-5 and 14-6

Results from the data compilation of Yaws and co-workers (44) were selected for heat capacity of ideal gas. Correlation of data was accomplished using a series expansion in temperature, Equation (1-5). Results are in favorable agreement with data. Errors are about 1% or less in most cases.

Results from the data compilation of Yaws and co-workers (44) were selected for heat capacity of liquid. The coverage applies to temperatures from below the boiling point to temperatures above the boiling point for most of the compounds. Data were correlated with a series expansion in temperature, Equation (1-6). Results are in favorable agreement with data.

VISCOSITY - Figures 14-7 and 14-8

The DIPPR project (5) was selected for viscosity of gas. Data for gas viscosity as a function of temperature were correlated using Equation (1-7). Results are in favorable agreement with data. Errors are about 1-10% or less in most cases.

The DIPPR project (5) was also selected for viscosity of liquid. Temperatures from below the boiling point to temperatures above the boiling point are covered for most of the compounds. Data for liquid viscosity as a function of temperature were correlated using the de Guzman - Andrade equation with extended terms, Equation (1-8). Correlation results and data are in favorable agreement with errors being about 1-5% or less.

THERMAL CONDUCTIVITY - Figures 14-9 and 14-10

Results from the DIPPR project (5) were selected for thermal conductivity of gas. Data for gas thermal conductivity as a function of temperature were correlated using the Equation (1-9). Reliability of results is good with errors of about 1-5% or less in most cases.

Results from the DIPPR project (5) were selected for thermal conductivity of liquid. The coverage applies to temperatures from below the boiling point to temperatures above the boiling point for most of the compounds. Data for liquid thermal conductivity as a function of temperature were correlated using a series expansion in temperature, Equation (1-10). Results are in favorable agreement with data. Errors are about 1-5% or less in most cases.

ENTHALPY OF FORMATION - Figure 14-11

The data compilation of Yaws and co-workers (44,45) was selected for enthalpy of formation of ideal gas. Data for enthalpy of formation of the ideal gas is a series expansion in temperature, Equation (1-11). Results from the correlation are in favorable agreement with data.

GIBB'S FREE ENERGY OF FORMATION - Figure 14-12

Results from the data compilation of Yaws and co-workers (44,46) were selected for Gibb's free energy of formation of ideal gas. Data for Gibb's free energy of formation of the ideal gas is a series expansion in temperature, Equation (1-12). Correlation results are in favorable agreement with data.

Table 14-1 Physical Properties

	n-Pentane	n-Hexane	n-Heptane	n-Octane
1. Name	n-Pentane	n-Hexane	n-Heptane	n-Octane
2. Formula	C5H12	C6H14	C7H16	C8H18
3. Molecular Weight, g/mol	72.150	86.177	100.203	114.23
4. Critical Temperature, K	469.65	507.43	540.26	568.83
5. Critical Pressure, bar	33.688	30.123	27.358	24.863
6. Critical Volume, ml/mol	312.34	370.13	432.25	492.05
7. Critical Compressibility Factor	0.269	0.264	0.263	0.259
8. Acentric Factor	0.2486	0.3047	0.3494	0.3962
9. Melting Point, K	143.42	177.83	182.57	216.38
10. Boiling Point @ 1 atm, K	309.22	341.88	371.58	398.83
11. Heat of Vaporization @ BP, kJ/kg	357.20	334.83	316.35	301.24
12. Density of Liquid @ 25 C, g/ml	0.620	0.654	0.680	0.699
13. Surface Tension @ 25 C, dynes/cm	15.50	17.88	19.65	21.15
14. Heat Capacity of Gas @ 25 C, J/g K	1.664	1.659	1.655	1.652
15. Heat Capacity of Liquid @ 25 C, J/g K	2.331	2.267	2.232	2.209
16. Viscosity of Gas @ 25 C, micropoise	70.25	64.70	57.57	54.96
17. Viscosity of Liquid @ 25 C, centipoise	0.226	0.294	0.384	0.509
18. Thermal Conductivity of Gas @ 25 C, W/m K	0.015	0.013	0.013	0.012
19. Thermal Conductivity of Liquid @ 25 C, W/m K	0.115	0.120	0.123	0.128
20. Enthalpy of Formation of Gas @ 25 C, kJ/mol	-146.57	-167.34	-187.97	-208.66
21. Gibbs Free Energy of Formation of Gas @ 25 C, kJ/mol	-8.79	-0.75	7.45	15.79
22. Flash Point, K	233.15	251.48	269.26	286.48
23. Autoignition Temperature, K	533.15	507.04	495.93	493.15
24. Lower Explosion Limit in Air, vol %	1.4	1.2	1.2	1.0
25. Upper Explosion Limit in Air, vol %	7.8	7.5	6.7	6.5
26. Solubility in Water @ 25 C, ppm(wt)	38.5	13.31	2.24	0.431

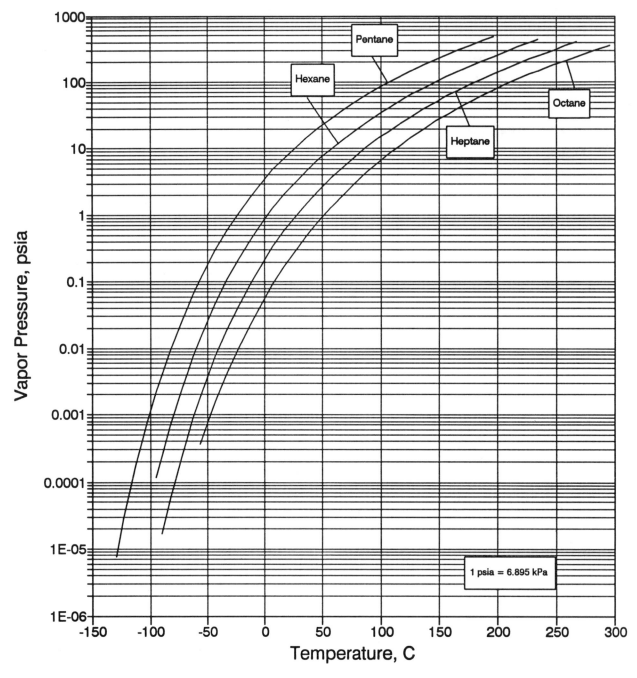

Figure 14-1 Vapor Pressure

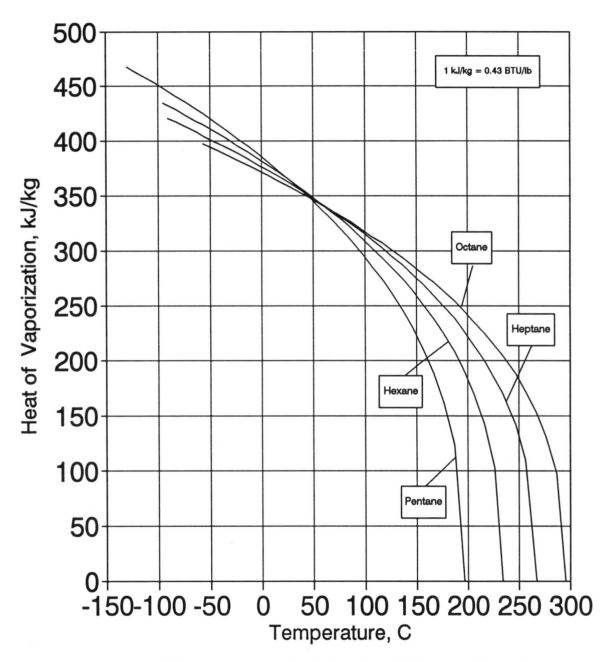

Figure 14-2 Heat of Vaporization

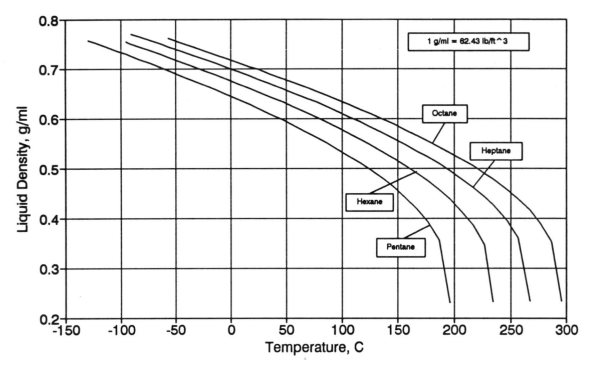

Figure 14-3 Liquid Density

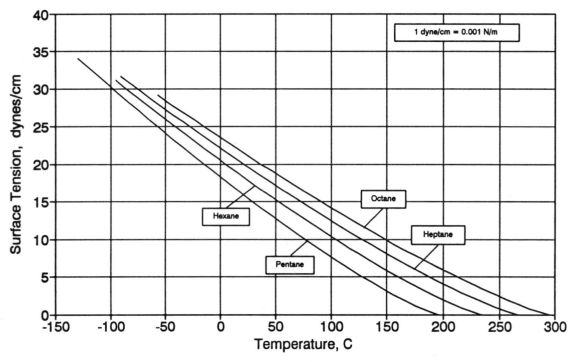

Figure 14-4 Surface Tension

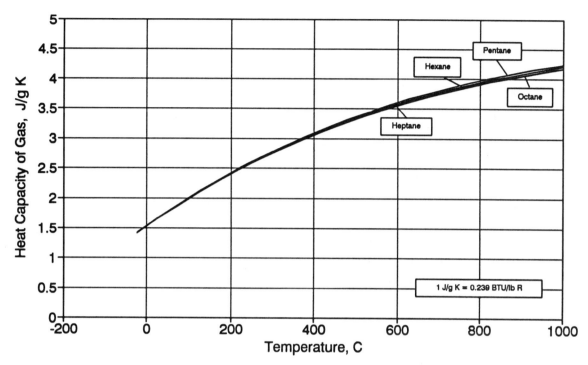

Figure 14-5 Heat Capacity of Gas

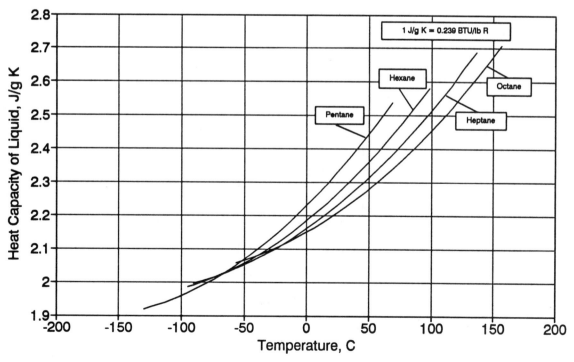

Figure 14-6 Heat Capacity of Liquid

161

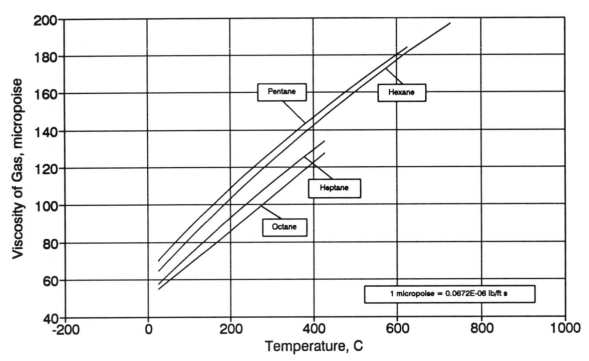

Figure 14-7 Viscosity of Gas

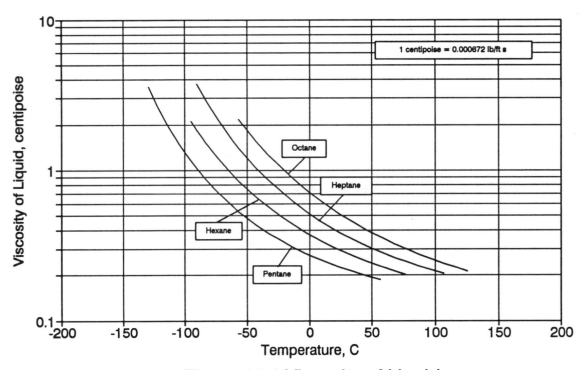

Figure 14-8 Viscosity of Liquid

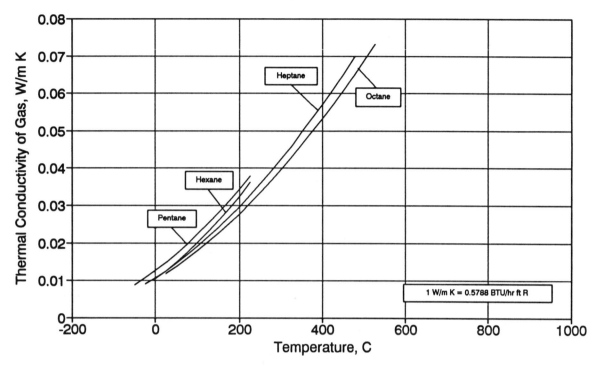

Figure 14-9 Thermal Conductivity of Gas

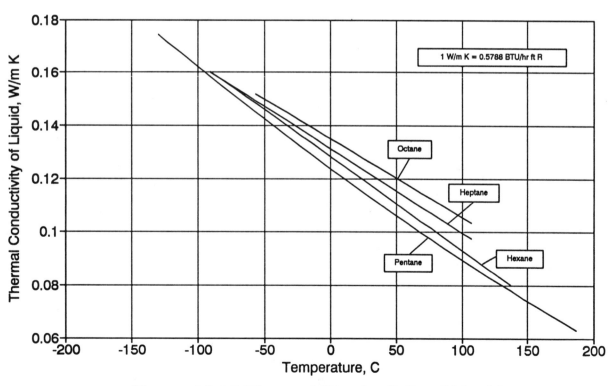

Figure 14-10 Thermal Conductivity of Liquid

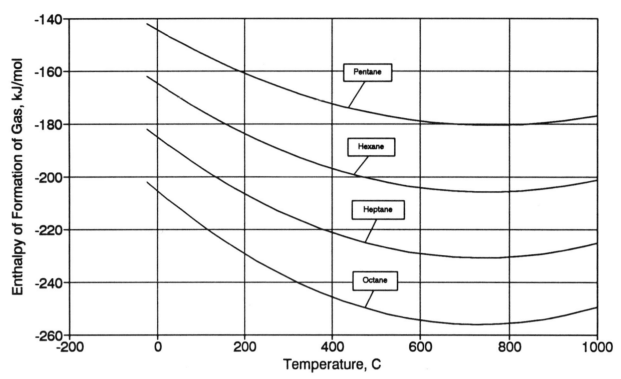

Figure 14-11 Enthalpy of Formation of Gas

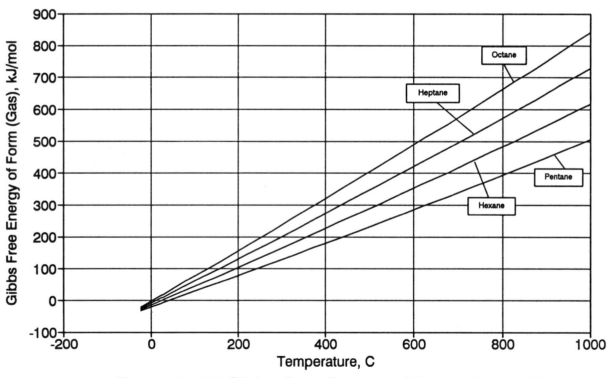

Figure 14-12 Gibbs Free Energy of Formation of Gas

Chapter 15

C_5 TO C_8 ALKENES

Robert W. Gallant and Carl L. Yaws

PHYSICAL PROPERTIES - Table 15-1

Property data from the literature (1-55,73,79,80,91,92,140,162,219-226) are given in Table 15-1. The critical constants have been determined experimentally and are available (1-7). Additional property data such as acentric factor, enthalpy of formation, lower explosion limit in air and solubility in water are also available. The DIPPR (Design Institute for Physical Property Research) project (5) and recent data compilations by Yaws and co-workers (44-55) were consulted extensively in preparing the tabulation.

VAPOR PRESSURE - Figure 15-1

Results from the DIPPR project (5) were selected for vapor pressure from very low temperatures to the critical point. Correlation of data for vapor pressure as a function of temperature was accomplished using Equation (1-1). Results from this equation (Antoine-type with extended terms) are in favorable agreement with experimental data. Errors are about 1-3% or less in most cases.

HEAT OF VAPORIZATION - Figure 15-2

The data compilation of Yaws and co-workers (44,52) was selected for heat of vaporization for temperatures ranging from melting point to critical point. The Watson equation, Equation (1-2), was used for correlation of the data as a function of temperature. Reliability of results is good with errors of about 1-5% or less.

LIQUID DENSITY - Figure 15-3

Results from the data compilation of Yaws and co-workers (44,54) were selected for liquid density from low temperatures at the melting point to higher temperatures up to the critical point. A modified Rackett equation, Equation (1-3), was used for correlation of the data as a function of temperature. Results from the correlation are in favorable agreement with data. Deviations are less than 1-5% in most cases.

SURFACE TENSION - Figure 15-4

The data compilation of Yaws and co-workers (44,55) was selected for surface tension for temperatures from melting point to critical point. Using data from the literature, correlation for surface tension as a function of temperature over the full liquid range was achieved by the modified Othmer equation, Equation (1-4). Accuracy is good with errors being about 1-3% or less in most cases.

HEAT CAPACITY - Figures 15-5 and 15-6

Results from the data compilation of Yaws and co-workers (44) were selected for heat capacity of ideal gas. Correlation of data was accomplished using a series expansion in temperature, Equation (1-5). Results are in favorable agreement with data. Errors are about 1% or less in most cases.

Results from the data compilation of Yaws and co-workers (44) were selected for heat capacity of liquid. The coverage applies to temperatures from below the boiling point to temperatures above the boiling point for most of the compounds. Data were correlated with a series expansion in temperature, Equation (1-6). Results are in favorable agreement with data.

VISCOSITY - Figures 15-7 and 15-8

The DIPPR project (5) was selected for viscosity of gas. Data for gas viscosity as a function of temperature were correlated using Equation (1-7). Results are in favorable agreement with data. Errors are about 1-5% or less in most cases.

The DIPPR project (5) was also selected for viscosity of liquid. Temperatures from below the boiling point to temperatures above the boiling point are covered for most of the compounds. Data for liquid viscosity as a function of temperature were correlated using the de Guzman - Andrade equation with extended terms, Equation (1-8). Correlation results and data are in agreement with errors being about 1-10% or less.

THERMAL CONDUCTIVITY - Figures 15-9 and 15-10

Results from the DIPPR project (5) were selected for thermal conductivity of gas. Data for gas thermal conductivity as a function of temperature were correlated using the Equation (1-9). Reliability of results is good with errors of about 1-10% or less in most cases.

Results from the DIPPR project (5) were selected for thermal conductivity of liquid. The coverage applies to temperatures from below the boiling point to temperatures above the boiling point for most of the compounds. Data for liquid thermal conductivity as a function of temperature were correlated using a series expansion in temperature, Equation (1-10). Results are in agreement with data. Errors are about 1-10% or less in most cases.

ENTHALPY OF FORMATION - Figure 15-11

The data compilation of Yaws and co-workers (44,45) was selected for enthalpy of formation of ideal gas. Data for enthalpy of formation of the ideal gas is a series expansion in temperature, Equation (1-11). Results from the correlation are in favorable agreement with data.

GIBB'S FREE ENERGY OF FORMATION - Figure 15-12

Results from the data compilation of Yaws and co-workers (44,46) were selected for Gibb's free energy of formation of ideal gas. Data for Gibb's free energy of formation of the ideal gas is a series expansion in temperature, Equation (1-12). Correlation results are in favorable agreement with data.

Table 15-1 Physical Properties

	1-Pentene	1-Hexene	1-Heptene	1-Octene
1. Name	1-Pentene	1-Hexene	1-Heptene	1-Octene
2. Formula	C5H10	C6H12	C7H14	C8H16
3. Molecular Weight, g/mol	70.134	84.161	98.188	112.214
4. Critical Temperature, K	464.78	504.03	537.29	566.60
5. Critical Pressure, bar	35.287	31.40	28.30	25.50
6. Critical Volume, ml/mol	296.00	354.00	413.00	472.00
7. Critical Compressibility Factor	0.270	0.265	0.262	0.256
8. Acentric Factor	0.233	0.280	0.331	0.3747
9. Melting Point, K	107.93	133.32	154.27	171.43
10. Boiling Point @ 1 atm, K	303.11	336.63	366.79	394.44
11. Heat of Vaporization @ BP, kJ/kg	361.55	337.04	316.61	301.40
12. Density of Liquid @ 25 C, g/ml	0.635	0.666	0.693	0.709
13. Surface Tension @ 25 C, dynes/cm	15.46	17.91	19.81	21.28
14. Heat Capacity of Gas @ 25 C, J/g K	1.561	1.571	1.580	1.586
15. Heat Capacity of Liquid @ 25 C, J/g K	2.211	2.181	2.152	2.126
16. Viscosity of Gas @ 25 C, micropoise	71.15	65.65	61.78	58.20
17. Viscosity of Liquid @ 25 C, centipoise	0.194	0.252	0.333	0.456
18. Thermal Conductivity of Gas @ 25 C, W/m K	0.0136	0.0133	0.0125	0.0116
19. Thermal Conductivity of Liquid @ 25 C, W/m K	0.116	0.121	0.124	0.128
20. Enthalpy of Formation of Gas @ 25 C, kJ/mol	-21.02	-41.78	-62.45	-83.09
21. Gibbs Free Energy of Formation of Gas @ 25 C, kJ/mol	78.80	87.05	95.37	103.69
22. Flash Point, K	255.37	-----	273.15	294.26
23. Autoignition Temperature, K	545.93	526.15	536.15	503.15
24. Lower Explosion Limit in Air, vol %	1.5	1.2	1.0	0.9
25. Upper Explosion Limit in Air, vol %	8.7	------	-----	-----
26. Solubility in Water @ 25 C, ppm(wt)	148.0	69.69	18.16	4.096

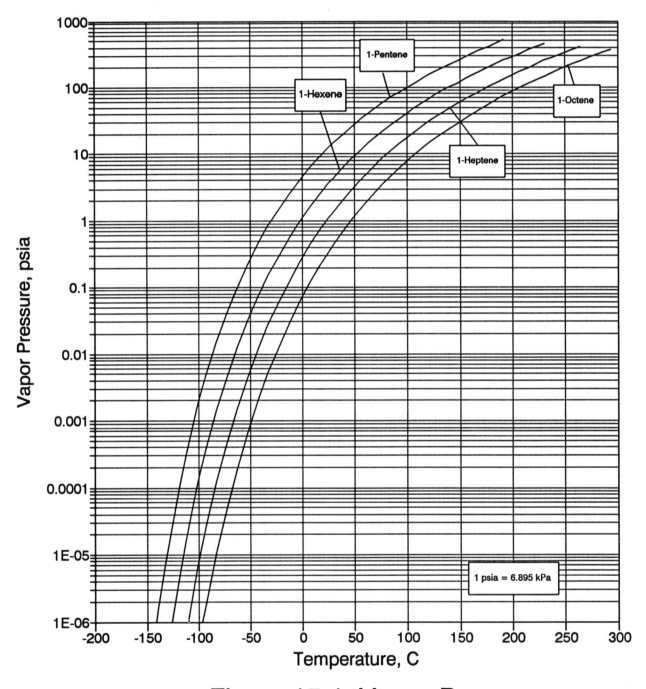

Figure 15-1 Vapor Pressure

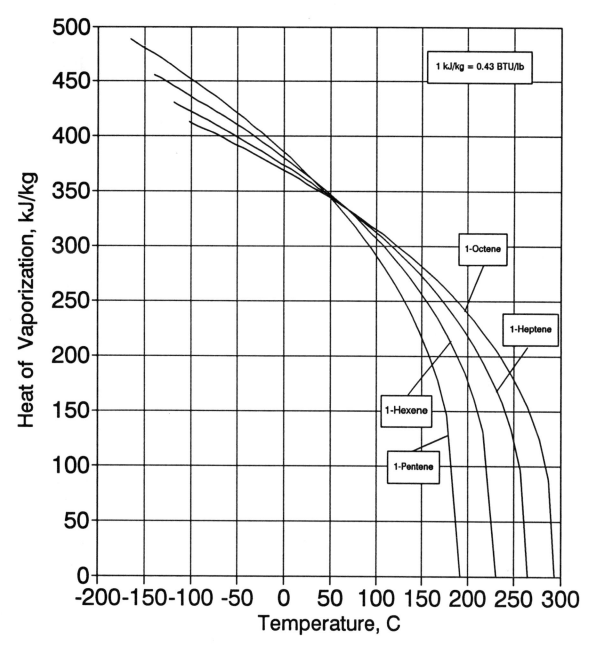

Figure 15-2 Heat of Vaporization

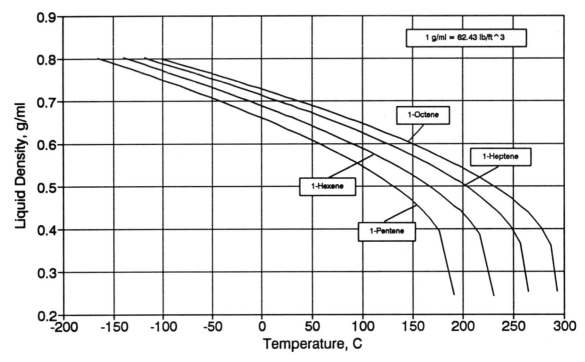

Figure 15-3 Liquid Density

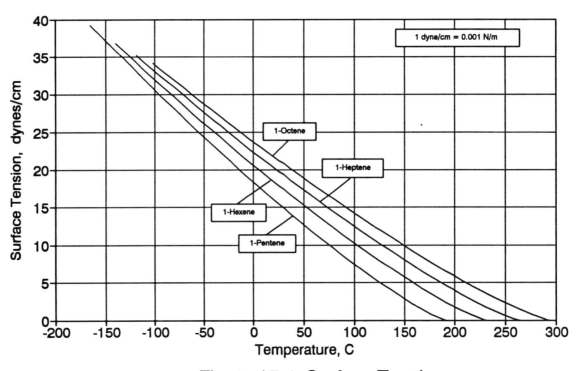

Figure 15-4 Surface Tension

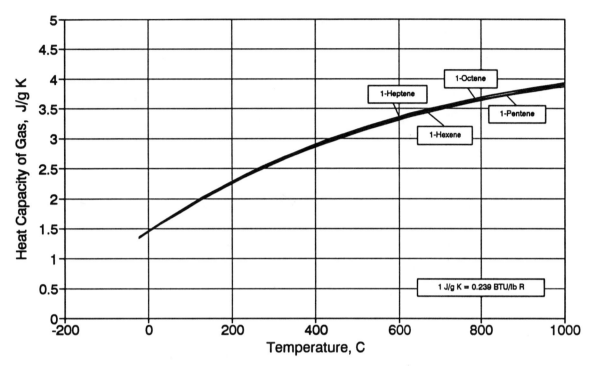

Figure 15-5 Heat Capacity of Gas

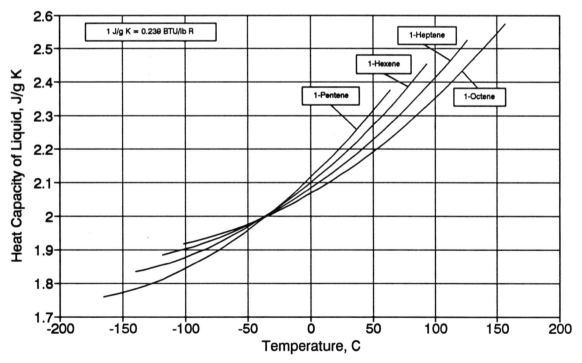

Figure 15-6 Heat Capacity of Liquid

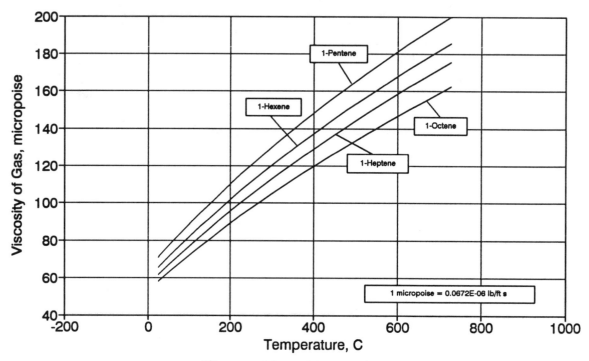

Figure 15-7 Viscosity of Gas

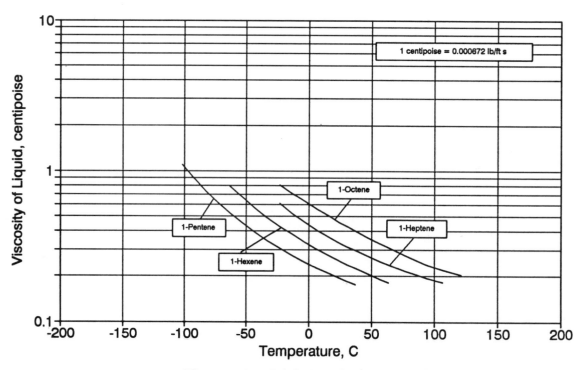

Figure 15-8 Viscosity of Liquid

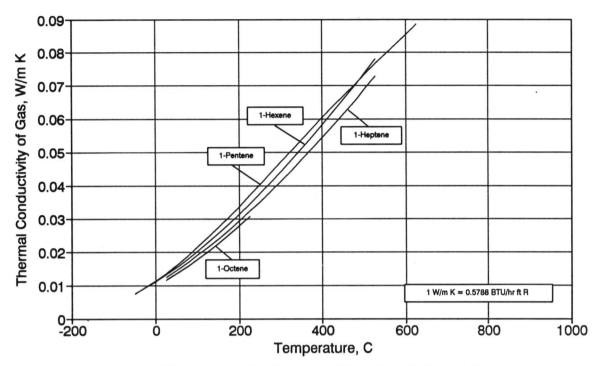

Figure 15-9 Thermal Conductivity of Gas

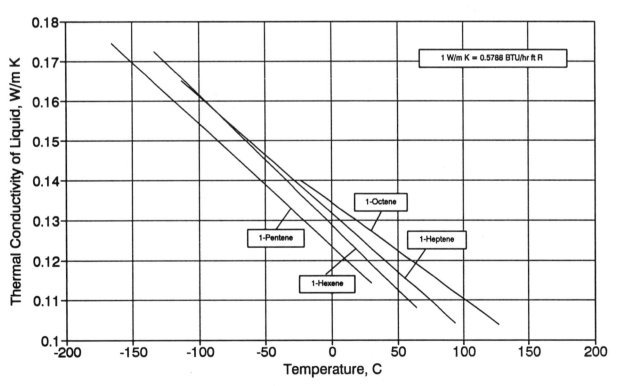

Figure 15-10 Thermal Conductivity of Liquid

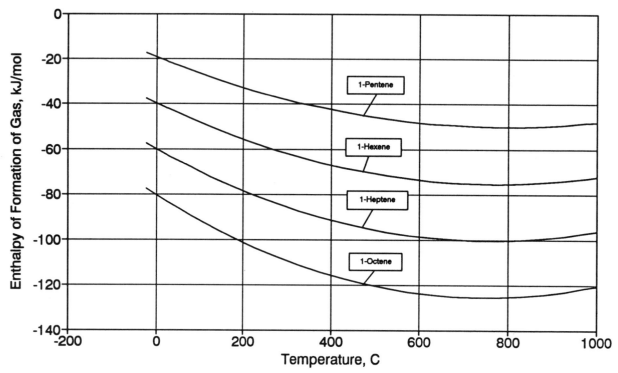

Figure 15-11 Enthalpy of Formation of Gas

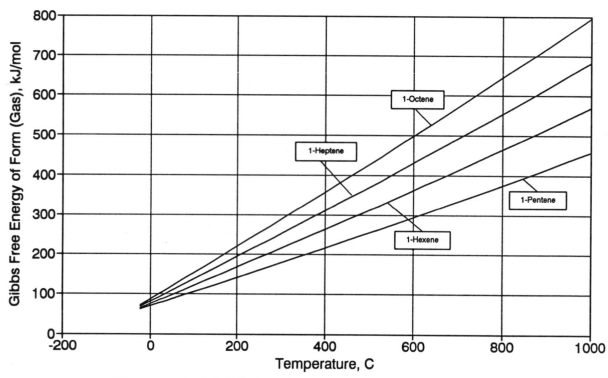

Figure 15-12 Gibbs Free Energy of Formation of Gas

Chapter 16

C₄ TO C₅ BRANCHED HYDROCARBONS

Robert W. Gallant and Carl L. Yaws

PHYSICAL PROPERTIES - Table 16-1

Property data from the literature (1-55,58,62,66,71,80,91,92,96,190,227-234) are given in Table 16-1. The critical constants are experimental values (1-7) except for isoprene (31,44) which are estimated values. Additional property data such as acentric factor, enthalpy of formation, lower explosion limit in air and solubility in water are also available. The DIPPR (Design Institute for Physical Property Research) project (5) and recent data compilations by Yaws and co-workers (44-55) were consulted extensively in preparing the tabulation.

VAPOR PRESSURE - Figure 16-1

Results from the DIPPR project (5) were selected for vapor pressure from very low temperatures to the critical point except for isoprene (31,44). Correlation of data for vapor pressure as a function of temperature was accomplished using Equation (1-1). Results from this equation (Antoine-type with extended terms) are in favorable agreement with experimental data. Errors are about 1-3% or less in most cases.

HEAT OF VAPORIZATION - Figure 16-2

The data compilation of Yaws and co-workers (44,52) was selected for heat of vaporization for temperatures ranging from melting point to critical point. The Watson equation, Equation (1-2), was used for correlation of the data as a function of temperature. Reliability of results is good with errors of about 1-5% or less.

LIQUID DENSITY - Figure 16-3

Results from the data compilation of Yaws and co-workers (44,54) were selected for liquid density from low temperatures at the melting point to higher temperatures up to the critical point. A modified Rackett equation, Equation (1-3), was used for correlation of the data as a function of temperature. Results from the correlation are in favorable agreement with data. Deviations are less than 1-5% in most cases.

SURFACE TENSION - Figure 16-4

The data compilation of Yaws and co-workers (44,55) was selected for surface tension for temperatures from melting point to critical point. Using data from the literature, correlation for surface tension as a function of temperature over the full liquid range was achieved by the modified Othmer equation, Equation (1-4). Accuracy is good with errors being about 1-3% or less in most cases.

HEAT CAPACITY - Figures 16-5 and 16-6

Results from the data compilation of Yaws and co-workers (44) were selected for heat capacity of ideal gas. Correlation of data was accomplished using a series expansion in temperature, Equation (1-5). Results are in favorable agreement with data. Errors are about 1% or less in most cases.

Results from the DIPPR project (5) were selected for heat capacity of liquid except for isoprene (31,44). The coverage applies to temperatures from below the boiling point to temperatures above the boiling point for most of the compounds. Data were correlated with a series expansion in temperature, Equation (1-6). Results are in favorable agreement with data.

VISCOSITY - Figures 16-7 and 16-8

The DIPPR project (5) was selected for viscosity of gas except for isoprene (31). Data for gas viscosity as a function of temperature were correlated using Equation (1-7). Results are in agreement with data. Errors are about 1-10% or less in most cases.

The DIPPR project (5) was also selected for viscosity of liquid except for isoprene (31). Temperatures from below the boiling point to temperatures above the boiling point are covered for most of the compounds. Data for liquid viscosity as a function of temperature were correlated using the de Guzman - Andrade equation with extended terms, Equation (1-8). Correlation results and data are in agreement with errors being about 1-10% or less except for isobutane (possible 25% error).

THERMAL CONDUCTIVITY - Figures 16-9 and 16-10

Results from the DIPPR project (5) were selected for thermal conductivity of gas except for isoprene (31). Data for gas thermal conductivity as a function of temperature were correlated using the Equation (1-9). Reliability of results is good with errors of about 1-10% or less in most cases.

Results from the DIPPR project (5) were selected for thermal conductivity of liquid except for isoprene (31). The coverage applies to temperatures from below the boiling point to temperatures above the boiling point for most of the compounds. Data for liquid thermal conductivity as a function of temperature were correlated using a series expansion in temperature, Equation (1-10). Results are in agreement with data. Errors are about 1-10% or less in most cases.

ENTHALPY OF FORMATION - Figure 16-11

The data compilation of Yaws and co-workers (44,45) was selected for enthalpy of formation of ideal gas. Data for enthalpy of formation of the ideal gas is a series expansion in temperature, Equation (1-11). Results from the correlation are in favorable agreement with data.

GIBB'S FREE ENERGY OF FORMATION - Figure 16-12

Results from the data compilation of Yaws and co-workers (44,46) were selected for Gibb's free energy of formation of ideal gas. Data for Gibb's free energy of formation of the ideal gas is a series expansion in temperature, Equation (1-12). Correlation results are in favorable agreement with data.

Table 16-1 Physical Properties

1. Name	Isobutane	Isobutylene	Isopentane	Isoprene (2-methyl-1,3-butadiene)
2. Formula	C_4H_{10}	C_4H_8	C_5H_{12}	C_5H_8
3. Molecular Weight, g/mol	58.123	56.107	72.150	68.118
4. Critical Temperature, K	408.14	417.90	460.43	483.30
5. Critical Pressure, bar	36.48	39.99	33.812	37.40
6. Critical Volume, ml/mol	262.70	238.88	305.83	266.00
7. Critical Compressibility Factor	0.282	0.275	0.270	0.248
8. Acentric Factor	0.177	0.1893	0.2275	0.164
9. Melting Point, K	113.54	132.81	113.25	127.20
10. Boiling Point @ 1 atm, K	261.43	266.25	300.99	307.22
11. Heat of Vaporization @ BP, kJ/kg	366.48	394.24	342.32	382.64
12. Density of Liquid @ 25 C, g/ml	0.551	0.588	0.614	0.675
13. Surface Tension @ 25 C, dynes/cm	9.76	11.68	14.45	16.14
14. Heat Capacity of Gas @ 25 C, J/g K	1.664	1.587	1.645	1.539
15. Heat Capacity of Liquid @ 25 C, J/g K	2.433	2.330	2.299	2.249
16. Viscosity of Gas @ 25 C, micropoise	76.14	81.36	72.20	77.30
17. Viscosity of Liquid @ 25 C, centipoise	0.166	0.153	0.214	0.202
18. Thermal Conductivity of Gas @ 25 C, W/m K	0.0165	0.0169	0.0148	0.0132
19. Thermal Conductivity of Liquid @ 25 C, W/m K	------	0.100	0.107	0.122
20. Enthalpy of Formation of Gas @ 25 C, kJ/mol	-134.65	-16.95	-154.63	75.62
21. Gibbs Free Energy of Formation of Gas @ 25 C, kJ/mol	-21.24	57.81	-15.24	145.64
22. Flash Point, K	------	-----	------	------
23. Autoignition Temperature, K	733.15	738.15	693.15	700.15
24. Lower Explosion Limit in Air, vol %	1.8	1.8	1.4	------
25. Upper Explosion Limit in Air, vol %	8.4	8.8	7.6	-----
26. Solubility in Water @ 25 C, ppm(wt)	48.90	263.0	47.80	642.0

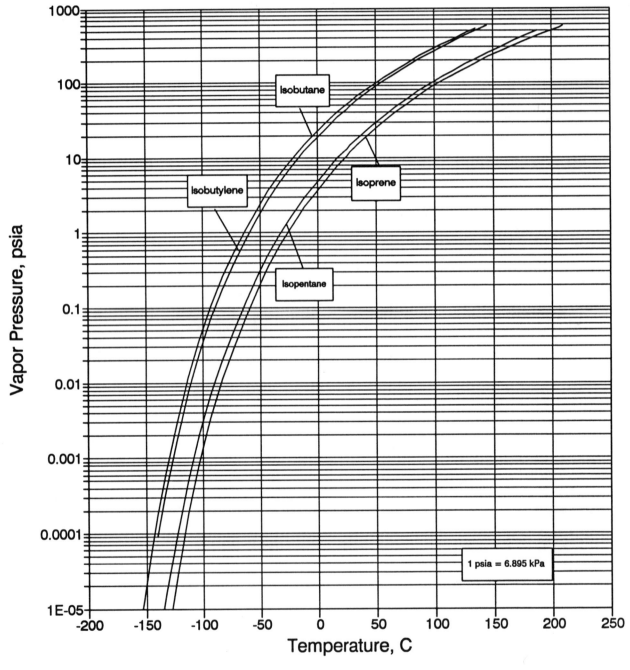

Figure 16-1 Vapor Pressure

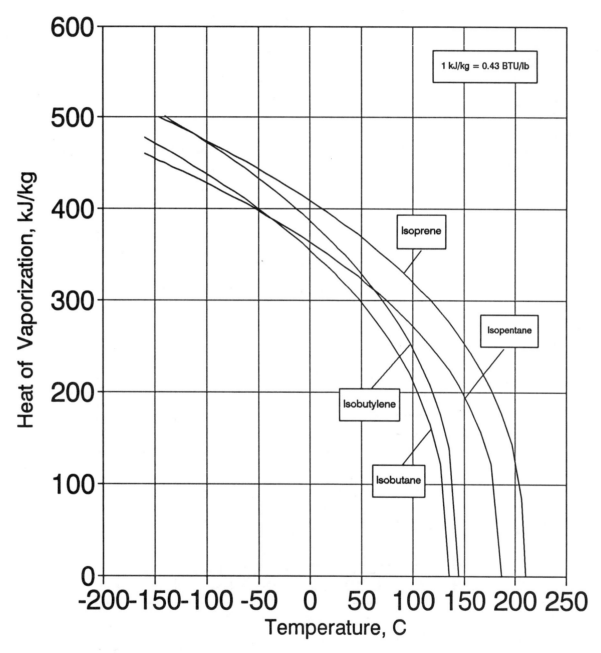

Figure 16-2 Heat of Vaporization

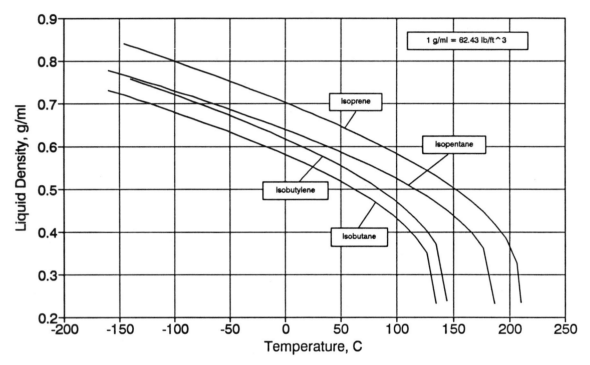

Figure 16-3 Liquid Density

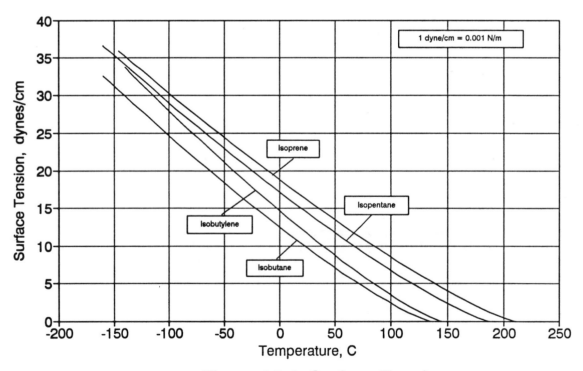

Figure 16-4 Surface Tension

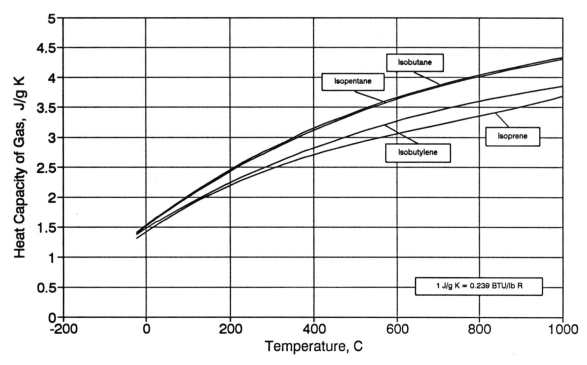

Figure 16-5 Heat Capacity of Gas

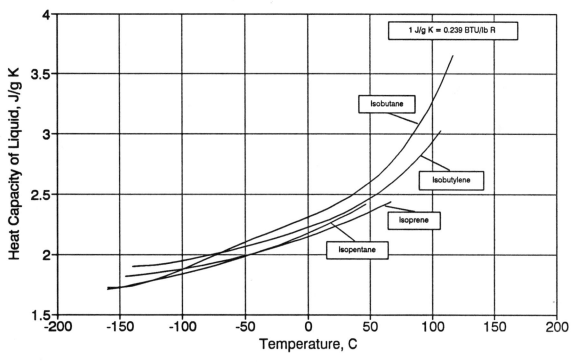

Figure 16-6 Heat Capacity of Liquid

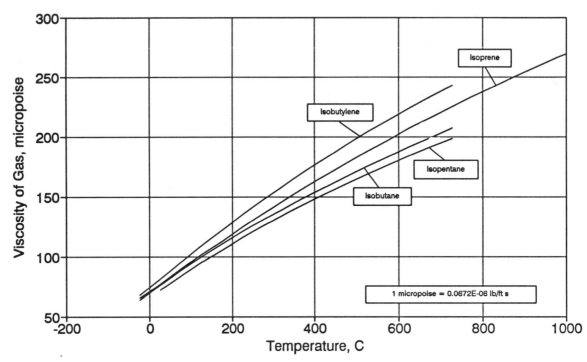

Figure 16-7 Viscosity of Gas

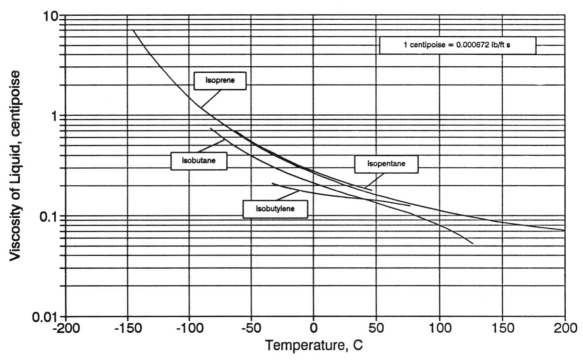

Figure 16-8 Viscosity of Liquid

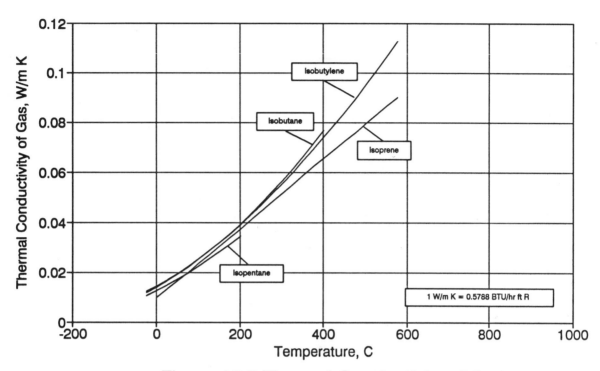

Figure 16-9 Thermal Conductivity of Gas

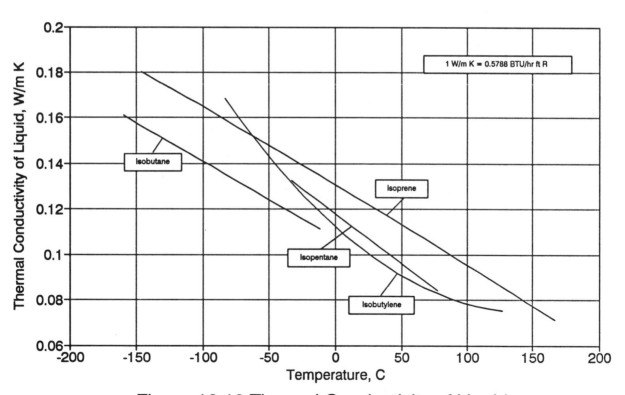

Figure 16-10 Thermal Conductivity of Liquid

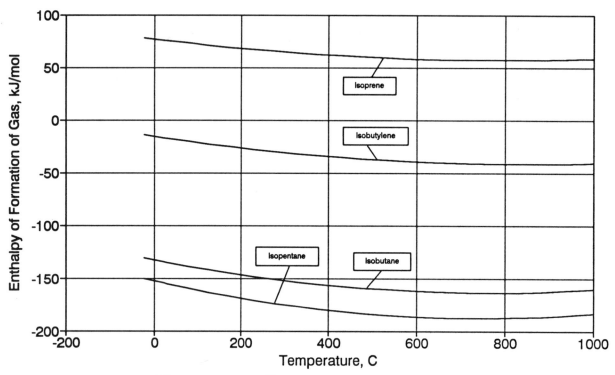

Figure 16-11 Enthalpy of Formation of Gas

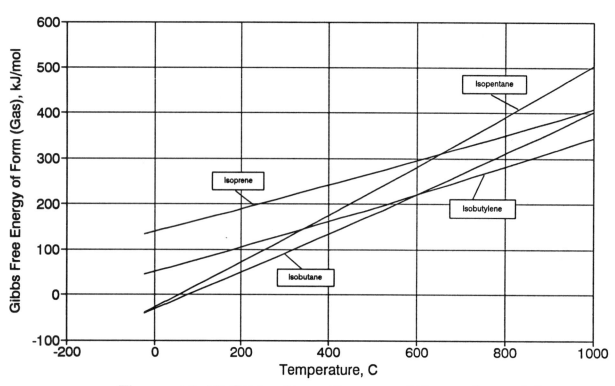

Figure 16-12 Gibbs Free Energy of Formation of Gas

Chapter 17

C_6 TO C_8 BRANCHED HYDROCARBONS

Robert W. Gallant and Carl L. Yaws

PHYSICAL PROPERTIES - Table 17-1

Property data from the literature (1-55,80,91-93,103,105,162,182,183,191,194,232,233,250-260) are given in Table 17-1. The critical constants are experimental values (1-7,44). Additional property data such as acentric factor, enthalpy of formation, lower explosion limit in air and solubility in water are also available. The DIPPR (Design Institute for Physical Property Research) project (5) and recent data compilations by Yaws and co-workers (44-55) were consulted extensively in preparing the tabulation.

VAPOR PRESSURE - Figure 17-1

Results from the DIPPR project (5) were selected for vapor pressure from very low temperatures to the critical point except for 2-methylheptane (44). Correlation of data for vapor pressure as a function of temperature was accomplished using Equation (1-1). Results from this equation (Antoine-type with extended terms) are in favorable agreement with experimental data. Errors are about 1-3% or less in most cases.

HEAT OF VAPORIZATION - Figure 17-2

The data compilation of Yaws and co-workers (44,52) was selected for heat of vaporization for temperatures ranging from melting point to critical point. The Watson equation, Equation (1-2), was used for correlation of the data as a function of temperature. Reliability of results is good with errors of about 1-5% or less.

LIQUID DENSITY - Figure 17-3

Results from the data compilation of Yaws and co-workers (44,54) were selected for liquid density from low temperatures at the melting point to higher temperatures up to the critical point. A modified Rackett equation, Equation (1-3), was used for correlation of the data as a function of temperature. Results from the correlation are in favorable agreement with data. Deviations are less than 1-5% in most cases.

SURFACE TENSION - Figure 17-4

The data compilation of Yaws and co-workers (44,55) was selected for surface tension for temperatures from melting point to critical point. Using data from the literature, correlation for surface tension as a function of temperature over the full liquid range was achieved by the modified Othmer equation, Equation (1-4). Accuracy is good with errors being about 1-3% or less in most cases.

HEAT CAPACITY - Figures 17-5 and 17-6

Results from the data compilation of Yaws and co-workers (44) were selected for heat capacity of ideal gas. Correlation of data was accomplished using a series expansion in temperature, Equation (1-5). Results are in favorable agreement with data. Errors are about 1% or less in most cases.

Results from the data compilation of Yaws and co-workers (44) were selected for heat capacity of liquid. The coverage applies to temperatures from below the boiling point to temperatures above the boiling point for most of the compounds. Data were correlated with a series expansion in temperature, Equation (1-6). Results are in favorable agreement with data.

VISCOSITY - Figures 17-7 and 17-8

The DIPPR project (5) was selected for viscosity of gas except for 2-methylheptane (30). Data for gas viscosity as a function of temperature were correlated using Equation (1-7). Results are in agreement with data. Errors are about 1-10% or less in most cases.

The DIPPR project (5) was also selected for viscosity of liquid except for 2-methylheptane (30). Temperatures from below the boiling point to temperatures above the boiling point are covered for most of the compounds. Data for liquid viscosity as a function of temperature were correlated using the de Guzman - Andrade equation with extended terms, Equation (1-8). Correlation results and data are in agreement with errors being about 1-10% or less.

THERMAL CONDUCTIVITY - Figures 17-9 and 17-10

Results from the DIPPR project (5) were selected for thermal conductivity of gas except for 2-methylheptane (30). Data for gas thermal conductivity as a function of temperature were correlated using the Equation (1-9). Reliability of results is fair with errors of about 5-25% or less.

Results from the DIPPR project (5) were selected for thermal conductivity of liquid. In the absence of data, values for 2-methylheptane (C_8H_{18}) were assumed to be equal to values for isooctane (C_8H_{18}). The coverage applies to temperatures from below the boiling point to temperatures above the boiling point for most of the compounds. Data for liquid thermal conductivity as a function of temperature were correlated using a series expansion in temperature, Equation (1-10). Results are in agreement with data. Errors are about 1-10% or less in most cases.

ENTHALPY OF FORMATION - Figure 17-11

The data compilation of Yaws and co-workers (44,45) was selected for enthalpy of formation of ideal gas. Data for enthalpy of formation of the ideal gas is a series expansion in temperature, Equation (1-11). Results from the correlation are in favorable agreement with data.

GIBB'S FREE ENERGY OF FORMATION - Figure 17-12

Results from the data compilation of Yaws and co-workers (44,46) were selected for Gibb's free energy of formation of ideal gas. Data for Gibb's free energy of formation of the ideal gas is a series expansion in temperature, Equation (1-12). Correlation results are in favorable agreement with data.

Table 17-1 Physical Properties

	2-Methyl-pentane	2-Methyl-hexane	Isooctane (2,2,4-trimethyl-pentane)	2-Methyl-heptane
1. Name	2-Methyl-pentane	2-Methyl-hexane	Isooctane (2,2,4-trimethyl-pentane)	2-Methyl-heptane
2. Formula	C_6H_{14}	C_7H_{16}	C_8H_{18}	C_8H_{18}
3. Molecular Weight, g/mol	86.177	100.203	114.230	114.230
4. Critical Temperature, K	497.50	530.37	543.96	560.60
5. Critical Pressure, bar	30.103	27.338	25.676	24.90
6. Critical Volume, ml/mol	366.37	421.00	468.00	488.30
7. Critical Compressibility Factor	0.267	0.261	0.266	0.261
8. Acentric Factor	0.2781	0.3282	0.3031	0.3780
9. Melting Point, K	119.55	154.90	165.78	164.00
10. Boiling Point @ 1 atm, K	333.41	363.20	372.39	390.80
11. Heat of Vaporization @ BP, kJ/kg	322.38	306.07	271.46	295.96
12. Density of Liquid @ 25 C, g/ml	0.648	0.674	0.688	0.694
13. Surface Tension @ 25 C, dynes/cm	16.88	18.81	18.33	20.27
14. Heat Capacity of Gas @ 25 C, J/g K	1.671	1.655	1.652	1.652
15. Heat Capacity of Liquid @ 25 C, J/g K	2.255	2.196	2.043	2.188
16. Viscosity of Gas @ 25 C, micropoise	65.61	61.73	62.47	55.50
17. Viscosity of Liquid @ 25 C, centipoise	0.281	0.360	0.477	0.475
18. Thermal Conductivity of Gas @ 25 C, W/m K	0.0125	0.0141	0.0133	0.0122
19. Thermal Conductivity of Liquid @ 25 C, W/m K	0.1065	0.1029	0.0998	------
20. Enthalpy of Formation of Gas @ 25 C, kJ/mol	-174.49	-195.12	-224.35	-215.69
21. Gibbs Free Energy of Formation of Gas @ 25 C, kJ/mol	-5.52	2.67	13.08	12.16
22. Flash Point, K	239.0	-----	260.93	------
23. Autoignition Temperature, K	579.26	553.15	690.93	------
24. Lower Explosion Limit in Air, vol %	1.2	1.0	1.1	------
25. Upper Explosion Limit in Air, vol %	7.0	6.0	6.0	-----
26. Solubility in Water @ 25 C, ppm(wt)	13.80	2.54	2.22	.85

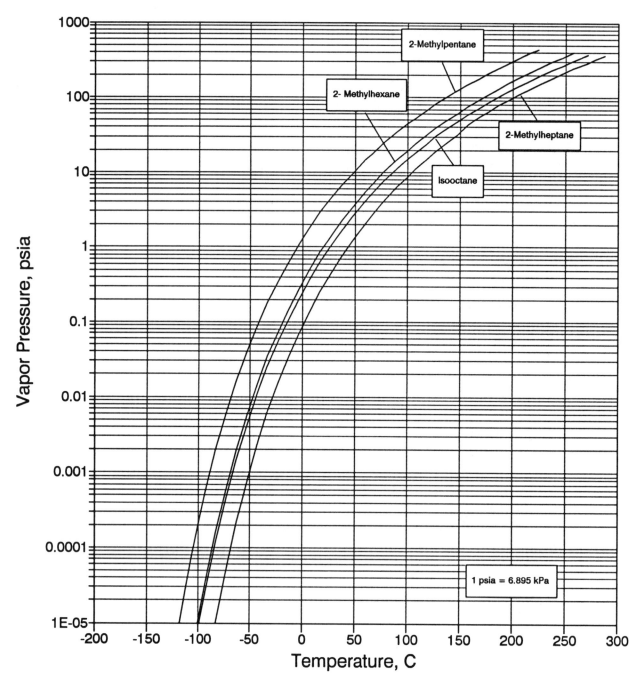

Figure 17-1 Vapor Pressure

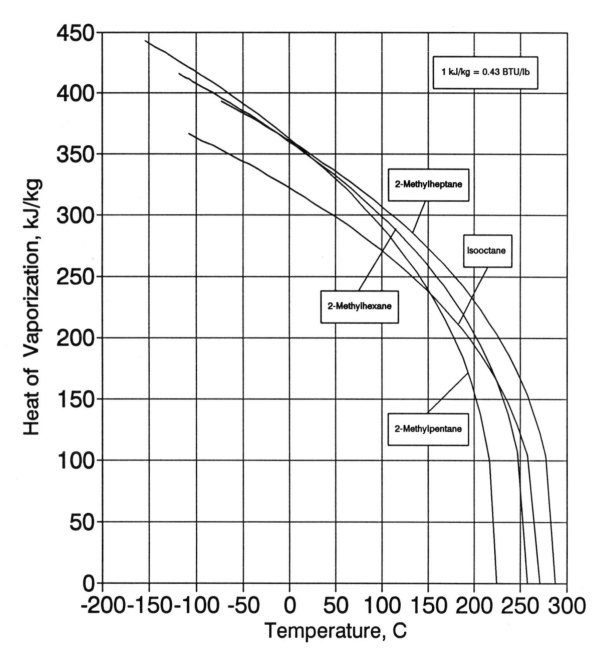

Figure 17-2 Heat of Vaporization

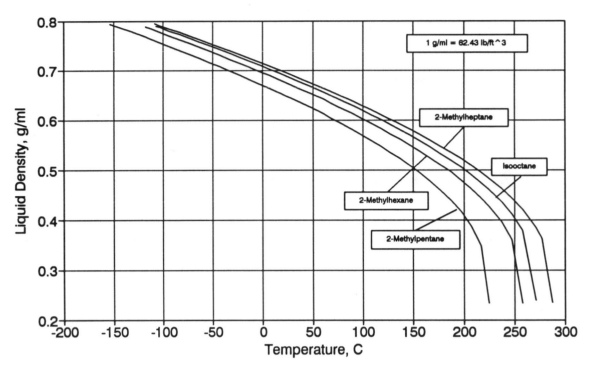

Figure 17-3 Liquid Density

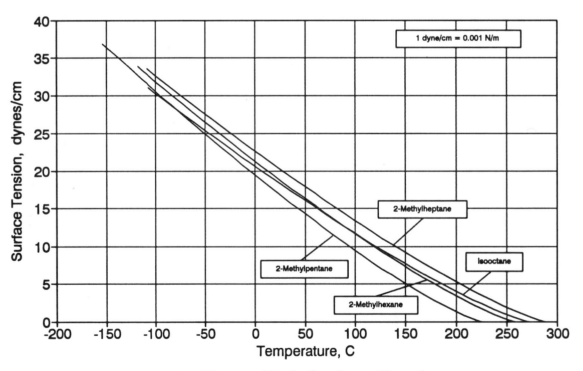

Figure 17-4 Surface Tension

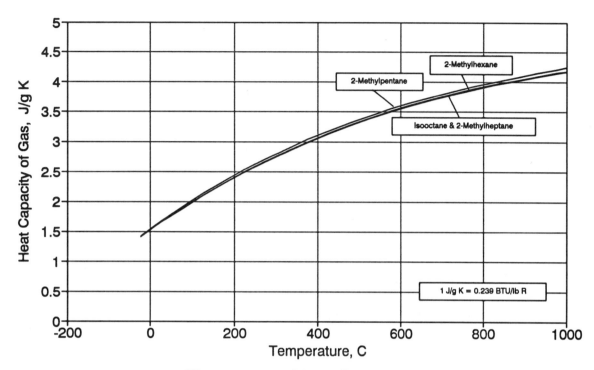

Figure 17-5 Heat Capacity of Gas

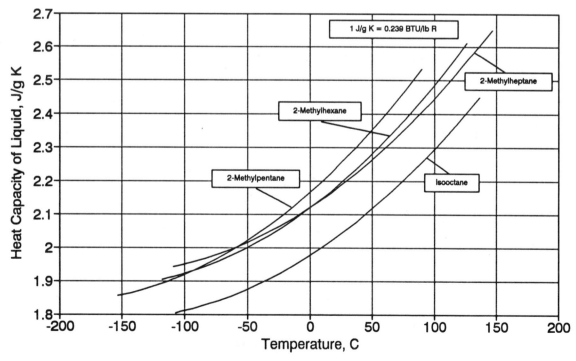

Figure 17-6 Heat Capacity of Liquid

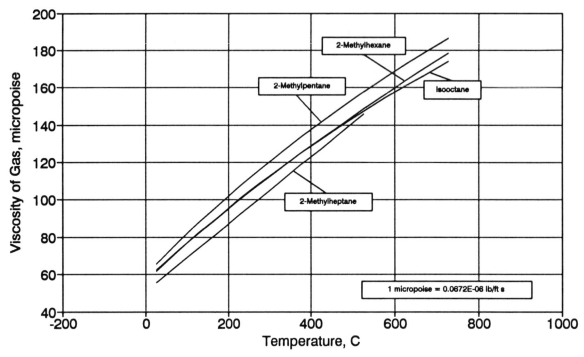

Figure 17-7 Viscosity of Gas

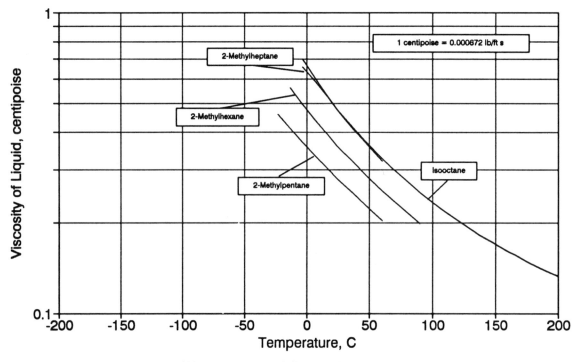

Figure 17-8 Viscosity of Liquid

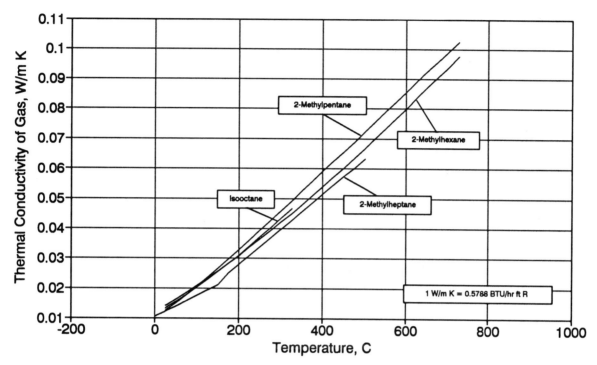

Figure 17-9 Thermal Conductivity of Gas

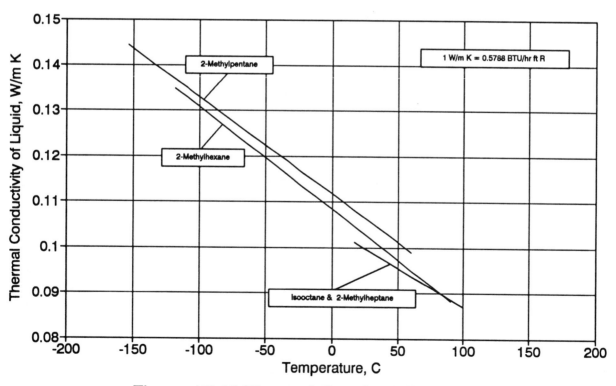

Figure 17-10 Thermal Conductivity of Liquid

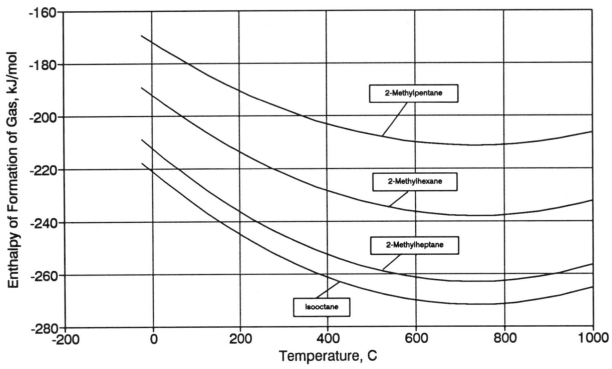

Figure 17-11 Enthalpy of Formation of Gas

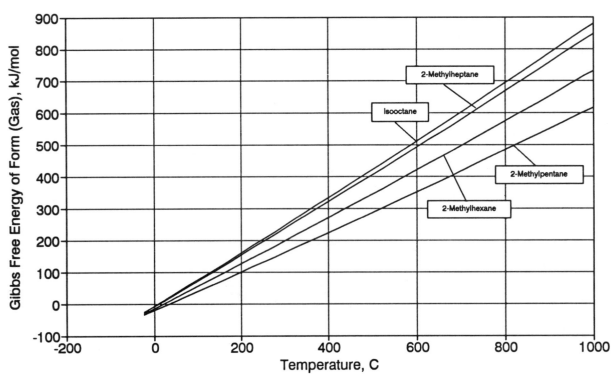

Figure 17-12 Gibbs Free Energy of Formation of Gas

Chapter 18

CHLORINATED C$_2$'s

Robert W. Gallant, Carl L. Yaws and Xiang Pan

PHYSICAL PROPERTIES - Table 18-1

Property data from the literature (1-55,88,89,100,105,107,111,113,114,117,125,139,261-263) are given in Table 18-1. The critical constants are from the DIPPR project (5) except for cis- and trans-dichloroethylene (44,47). Additional property data such as acentric factor, enthalpy of formation, lower explosion limit in air and solubility in water are also available. The DIPPR (Design Institute for Physical Property Research) project (5) and recent data compilations by Yaws and co-workers (44-55) were consulted extensively in preparing the tabulation.

VAPOR PRESSURE - Figure 18-1

Results from the DIPPR project (5) were selected for vapor pressure from very low temperatures to the critical point except for cis- and trans-dichloroethylene (44,48). Correlation of data for vapor pressure as a function of temperature was accomplished using Equation (1-1). Results from this equation (Antoine-type with extended terms) are in favorable agreement with experimental data. Errors are about 1-3% or less in most cases.

HEAT OF VAPORIZATION - Figure 18-2

Results from the DIPPR project (5) were selected for heat of vaporization for temperatures ranging from melting point to critical point except for cis- and trans-dichloroethylene (44,52). The Watson equation, Equation (1-2a), was used for correlation of the data as a function of temperature. Reliability of results is good with errors of about 1-5% or less.

LIQUID DENSITY - Figure 18-3

Results from the DIPPR project (5) were selected for liquid density from low temperatures at the melting point to higher temperatures up to the critical point except for cis- and trans-dichloroethylene (44,54). A modified Rackett equation, Equation (1-3a), was used for correlation of the data as a function of temperature. Results from the correlation are in favorable agreement with data. Deviations are less than 1-5% in most cases.

SURFACE TENSION - Figure 18-4

The DIPPR project (5) was selected for surface tension for temperatures from melting point to critical point except for cis- and trans-dichloroethylene (44,55). Using data from the literature, correlation for surface tension as a function of temperature over the full liquid range was achieved by the modified Othmer equation, Equation (1-4a). Accuracy is good with errors being about 1-3% or less in most cases.

HEAT CAPACITY - Figures 18-5 and 18-6

Results from the DIPPR project (5) were selected for heat capacity of ideal gas except for cis- and trans-dichloroethylene (44). Correlation of data was accomplished using Equation (1-5a). Results are in favorable agreement with data. Errors are about 1% or less in most cases.

Results from the data compilation of Yaws and co-workers (44) were selected for heat capacity of liquid except for methylchloroform (30). The coverage applies to temperatures from below the boiling point to temperatures above the boiling point for most of the compounds. Data were correlated with a series expansion in temperature, Equation (1-6). Results are in favorable agreement with data.

VISCOSITY - Figures 18-7 and 18-8

The DIPPR project (5) was selected for viscosity of gas except for cis- and trans-dichloroethylene (30). Data for gas viscosity as a function of temperature were correlated using Equation (1-7). Results are in agreement with data. Errors are about 1-10% or less in most cases.

The DIPPR project (5) was also selected for viscosity of liquid except for cis- and trans-dichloroethylene (30). Temperatures from below the boiling point to temperatures above the boiling point are covered for most of the compounds. Data for liquid viscosity as a function of temperature were correlated using the de Guzman - Andrade equation with extended terms, Equation (1-8). Correlation results and data are in agreement with errors being about 1-10% or less.

THERMAL CONDUCTIVITY - Figures 18-9 and 18-10

Results from the DIPPR project (5) were selected for thermal conductivity of gas except for cis- and trans-dichloroethylene (30). Data for gas thermal conductivity as a function of temperature were correlated using the Equation (1-9). Reliability of results is good with errors of about 1-10% or less.

In the absence of data and prediction equations with high accuracy for polar compounds, values for cis- and trans-dichloroethylene ($C_2H_2Cl_2$) were assumed to be intermediate to values for trichloroethylene (C_2HCl_3) and chloroethylene (C_2H_3Cl). For the other compounds, results from the DIPPR project (5) were selected for thermal conductivity of liquid. The coverage applies to temperatures from below the boiling point to temperatures above the boiling point for most of the compounds. Data for liquid thermal conductivity as a function of temperature were correlated using a series expansion in temperature, Equation (1-10). Results are in agreement with data. Errors are about 1-10% or less in most cases. The estimates for cis- and trans-dichloroethylene should be considered rough values.

ENTHALPY OF FORMATION - Figure 18-11

The data compilation of Yaws and co-workers (44,45) was selected for enthalpy of formation of ideal gas. Data for enthalpy of formation of the ideal gas is a series expansion in temperature, Equation (1-11). Results from the correlation are in favorable agreement with data.

GIBB'S FREE ENERGY OF FORMATION - Figure 18-12

Results from the data compilation of Yaws and co-workers (44,46) were selected for Gibb's free energy of formation of ideal gas. Data for Gibb's free energy of formation of the ideal gas is a series expansion in temperature, Equation (1-12). Correlation results are in favorable agreement with data.

Table 18-1 Physical Properties

	Trans-1,2-dichloro-ethylene	Cis-1,2-dichloro-ethylene	Methyl-chloroform	1,1,2,2-Tetra-chloroethane
1. Name				
2. Formula	$C_2H_2Cl_2$	$C_2H_2Cl_2$	$C_2H_3Cl_3$	$C_2H_2Cl_4$
3. Molecular Weight, g/mol	96.944	96.944	133.405	167.850
4. Critical Temperature, K	513.00	537.00	545.00	645.00
5. Critical Pressure, bar	45.800	58.700	42.962	40.900
6. Critical Volume, ml/mol	225.50	225.50	281.00	325.00
7. Critical Compressibility Factor	0.258	0.249	0.266	0.248
8. Acentric Factor	0.2320	0.2380	0.2157	0.2592
9. Melting Point, K	223.00	192.70	242.75	229.35
10. Boiling Point @ 1 atm, K	321.50	333.50	347.23	418.25
11. Heat of Vaporization @ BP, kJ/kg	312.60	320.28	223.67	247.60
12. Density of Liquid @ 25 C, g/ml	1.2460	1.2818	1.3303	1.5776
13. Surface Tension @ 25 C, dynes/cm	23.05	23.11	25.02	33.91
14. Heat Capacity of Gas @ 25 C, J/g K	0.687	0.671	0.695	0.601
15. Heat Capacity of Liquid @ 25 C, J/g K	0.88	0.90	1.03	0.84
16. Viscosity of Gas @ 25 C, micropoise	97.00	97.00	------	------
17. Viscosity of Liquid @ 25 C, centipoise	0.390	0.390	0.790	1.625
18. Thermal Conductivity of Gas @ 25 C, W/m K	0.0051	0.0044	------	------
19. Thermal Conductivity of Liquid @ 25 C, W/m K	0.1058	0.1058	0.1012	0.1127
20. Enthalpy of Formation of Gas @ 25 C, kJ/mol	-0.42	-2.8	-142.30	-149.00
21. Gibbs Free Energy of Formation of Gas @ 25 C, kJ/mol	22.01	19.66	-76.19	-79.50
22. Flash Point, K	------	------	------	------
23. Autoignition Temperature, K	------	------	810.15	------
24. Lower Explosion Limit in Air, vol %	------	------	8.0	------
25. Upper Explosion Limit in Air, vol %	------	------	10.5	------
26. Solubility in Water @ 25 C, ppm(wt)	6300	3500	------	2900

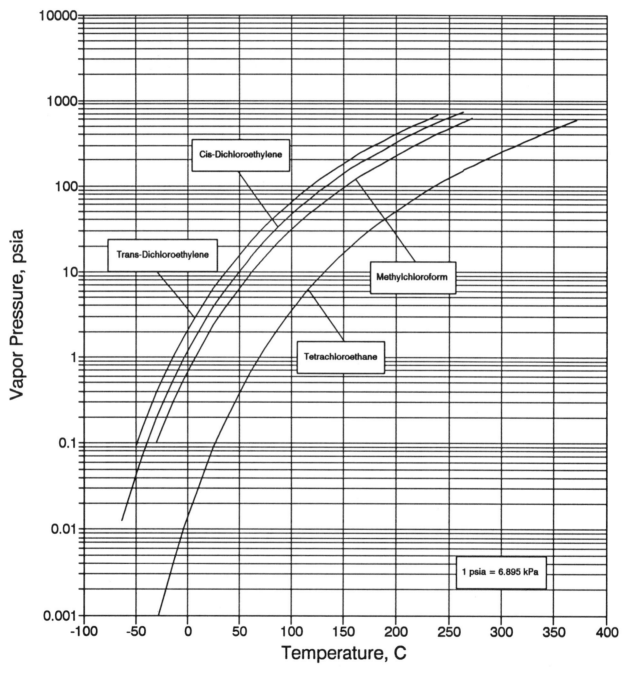

Figure 18-1 Vapor Pressure

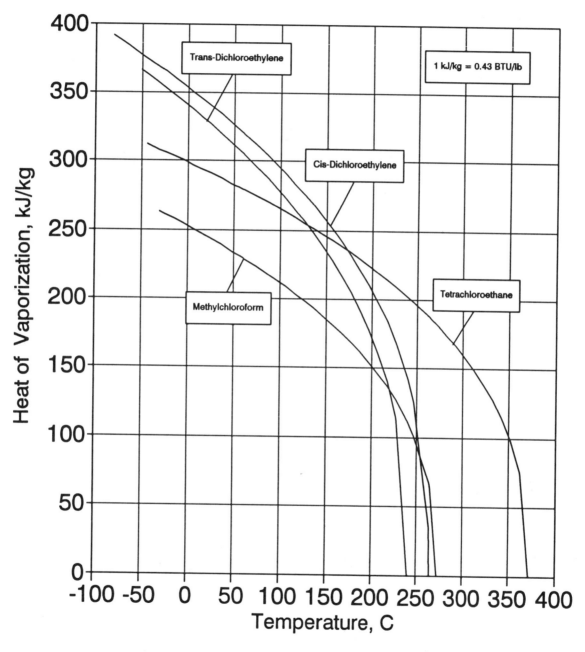

Figure 18-2 Heat of Vaporization

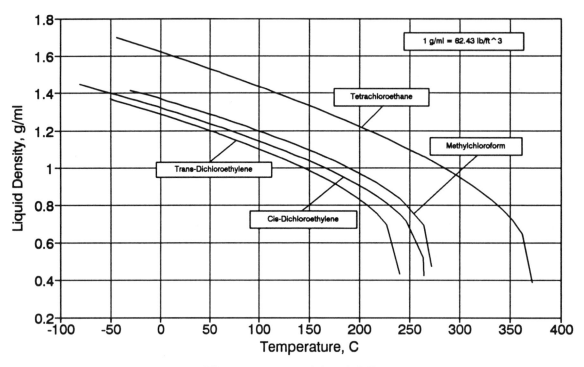

Figure 18-3 Liquid Density

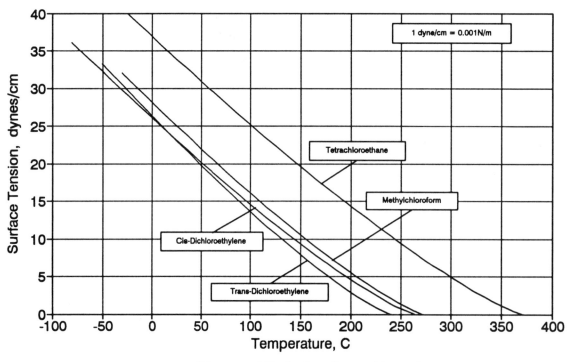

Figure 18-4 Surface Tension

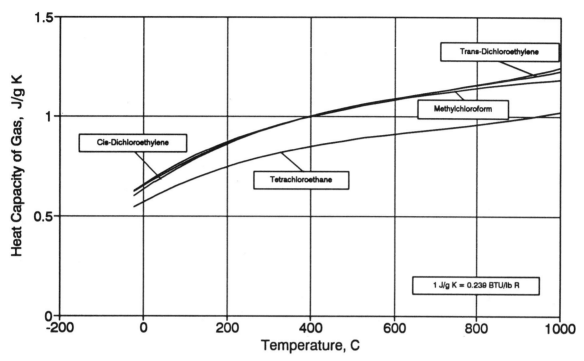

Figure 18-5 Heat Capacity of Gas

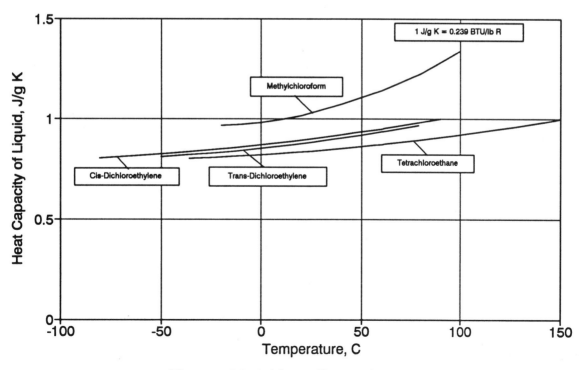

Figure 18-6 Heat Capacity of Liquid

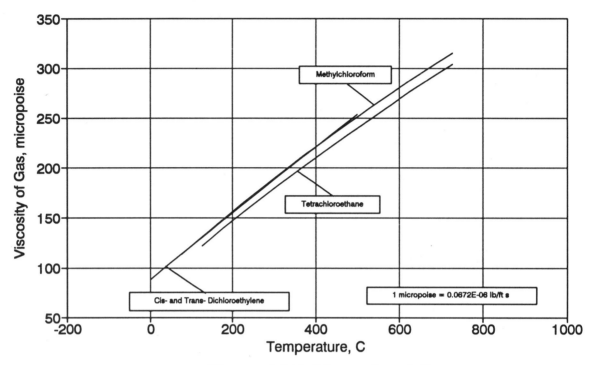

Figure 18-7 Viscosity of Gas

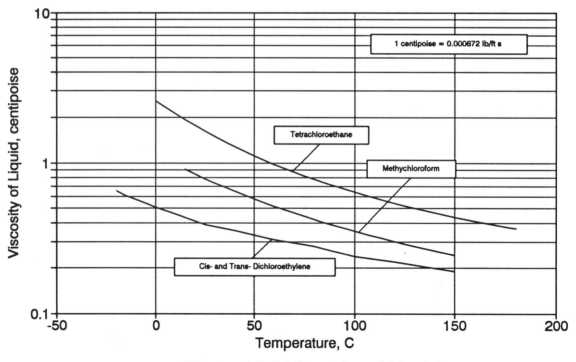

Figure 18-8 Viscosity of Liquid

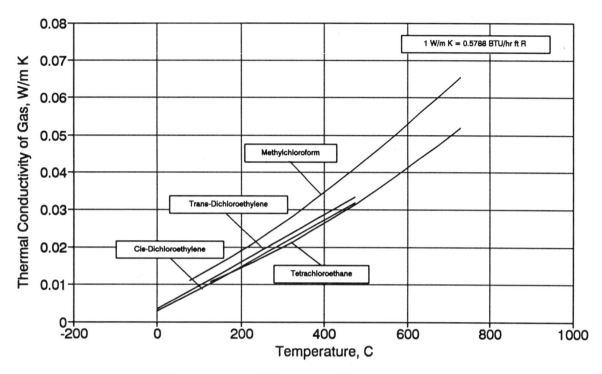

Figure 18-9 Thermal Conductivity of Gas

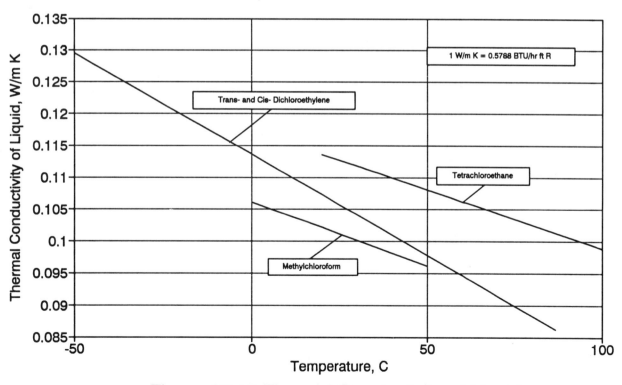

Figure 18-10 Thermal Conductivity of Liquid

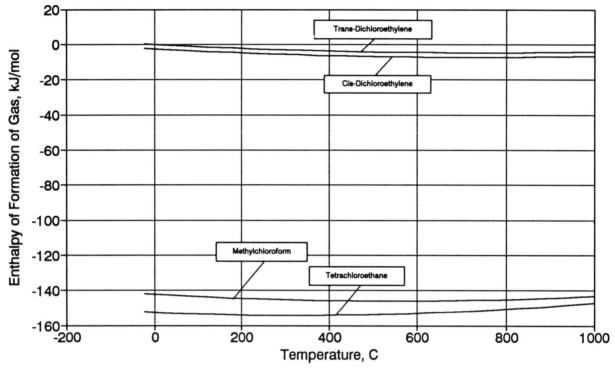

Figure 18-11 Enthalpy of Formation of Gas

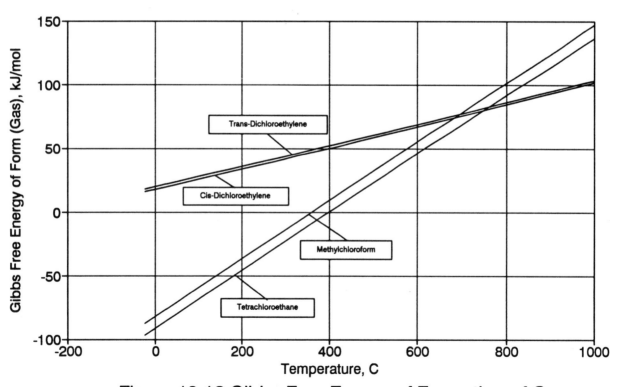

Figure 18-12 Gibbs Free Energy of Formation of Gas

Chapter 19

HALOGENATED METHANES

Robert W. Gallant, Carl L. Yaws, Jack R. Hopper and Duane G. Piper, Jr.

PHYSICAL PROPERTIES - Table 19-1

Property data from the literature (1-55,111,264-277) are given in Table 19-1. The critical constants are experimental values. Additional property data such as acentric factor, enthalpy of formation, lower explosion limit in air and solubility in water are also available. The DIPPR (Design Institute for Physical Property Research) project (5) and recent data compilations by Yaws and co-workers (44-55) were consulted extensively in preparing the tabulation.

VAPOR PRESSURE - Figure 19-1

Results from the DIPPR project (5) were selected for vapor pressure from very low temperatures to the critical point. Correlation of data for vapor pressure as a function of temperature was accomplished using Equation (1-1). Results from this equation (Antoine-type with extended terms) are in favorable agreement with experimental data. Errors are about 1-3% or less in most cases.

HEAT OF VAPORIZATION - Figure 19-2

The data compilation of Yaws and co-workers (44,52) was selected for heat of vaporization for temperatures ranging from melting point to critical point. The Watson equation, Equation (1-2), was used for correlation of the data as a function of temperature. Reliability of results is good with errors of about 1-5% or less.

LIQUID DENSITY - Figure 19-3

Results from the data compilation of Yaws and co-workers (44,54) were selected for liquid density from low temperatures at the melting point to higher temperatures up to the critical point. A modified Rackett equation, Equation (1-3), was used for correlation of the data as a function of temperature. Results from the correlation are in favorable agreement with data. Deviations are less than 1-5% in most cases.

SURFACE TENSION - Figure 19-4

The data compilation of Yaws and co-workers (44,55) was selected for surface tension for temperatures from melting point to critical point. Using data from the literature, correlation for surface tension as a function of temperature over the full liquid range was achieved by the modified Othmer equation, Equation (1-4). Accuracy is good with errors being about 1-3% or less in most cases.

HEAT CAPACITY - Figures 19-5 and 19-6

Results from the data compilation of Yaws and co-workers (44) were selected for heat capacity of ideal gas. Correlation of data was accomplished using a series expansion in temperature, Equation (1-5). Results are in favorable agreement with data. Errors are about 1-10% or less in most cases.

Results from the data compilation of Yaws and co-workers (44) were selected for heat capacity of liquid. The coverage applies to temperatures from below the boiling point to temperatures above the boiling point for most of the compounds. Data were correlated with a series expansion in temperature, Equation (1-6). Results are in favorable agreement with data.

VISCOSITY - Figures 19-7 and 19-8

The DIPPR project (5) was selected for viscosity of gas. Data for gas viscosity as a function of temperature were correlated using Equation (1-7). Results are in good agreement with data. Errors are about 1-10% or less in most cases.

The DIPPR project (5) was also selected for viscosity of liquid. Temperatures from below the boiling point to temperatures above the boiling point are covered for most of the compounds. Data for liquid viscosity as a function of temperature were correlated using the de Guzman - Andrade equation with extended terms, Equation (1-8). Correlation results and data are in agreement with errors being about 5-10% or less.

THERMAL CONDUCTIVITY - Figures 19-9 and 19-10

Results from the DIPPR project (5) were selected for thermal conductivity of gas. Data for gas thermal conductivity as a function of temperature were correlated using the Equation (1-9). Reliability of results is good with errors of about 1-5% or less.

Results from the DIPPR project (5) were selected for thermal conductivity of liquid. The coverage applies to temperatures from below the boiling point to temperatures above the boiling point for most of the compounds. Data for liquid thermal conductivity as a function of temperature were correlated using a series expansion in temperature, Equation (1-10). Results are in favorable agreement with data. Errors are about 1-5% or less in most cases.

ENTHALPY OF FORMATION - Figure 19-11

The data compilation of Yaws and co-workers (44,45) was selected for enthalpy of formation of ideal gas. Data for enthalpy of formation of the ideal gas is a series expansion in temperature, Equation (1-11). Results from the correlation are in favorable agreement with data.

GIBB'S FREE ENERGY OF FORMATION - Figure 19-12

Results from the data compilation of Yaws and co-workers (44,46) were selected for Gibb's free energy of formation of ideal gas. Data for Gibb's free energy of formation of the ideal gas is a series expansion in temperature, Equation (1-12). Correlation results are in favorable agreement with data.

Table 19-1 Physical Properties

1. Name	Fluorocarbon-14 Tetrafluoro-methane	Fluorocarbon-13 Chlorotrifluoro-methane	Fluorocarbon-12 Dichlorodifluoro-methane	Fluorocarbon-11 Trichlorofluoro-methane
2. Formula	CF_4	$CClF_3$	CCl_2F_2	CCl_3F
3. Molecular Weight, g/mol	88.005	104.46	120.914	137.368
4. Critical Temperature, K	227.50	301.96	384.95	471.20
5. Critical Pressure, bar	37.389	39.460	41.249	44.076
6. Critical Volume, ml/mol	140.00	180.28	217.00	248.00
7. Critical Compressibility Factor	0.277	0.283	0.280	0.279
8. Acentric Factor	0.1855	0.180	0.1796	0.1837
9. Melting Point, K	89.56	92.150	115.15	162.04
10. Boiling Point @ 1 atm, K	145.09	191.74	243.36	296.97
11. Heat of Vaporization @ BP, kJ/kg	136.04	148.42	165.10	180.36
12. Density of Liquid @ 25 C, g/ml	------	0.839	1.339	1.475
13. Surface Tension @ 25 C, dynes/cm	------	0.31	9.10	18.31
14. Heat Capacity of Gas @ 25 C, J/g K	0.698	0.640	0.600	0.569
15. Heat Capacity of Liquid @ 25 C, J/g K	------	------	0.957	0.887
16. Viscosity of Gas @ 25 C, micropoise	173.01	144.13	125.00	113.98
17. Viscosity of Liquid @ 25 C, centipoise	------	------	0.222	0.425
18. Thermal Conductivity of Gas @ 25 C, W/m K	0.0149	0.0122	0.0096	0.0079
19. Thermal Conductivity of Liquid @ 25 C, W/m K	------	0.042	0.070	0.091
20. Enthalpy of Formation of Gas @ 25 C, kJ/mol	-933.09	-707.98	-493.33	-284.99
21. Gibbs Free Energy of Formation of Gas @ 25 C, kJ/mol	-888.48	-653.88	-438.27	-245.03
22. Flash Point, K	------	------	------	------
23. Autoignition Temperature, K	------	------	------	------
24. Lower Explosion Limit in Air, vol %	------	------	------	------
25. Upper Explosion Limit in Air, vol %	------	------	------	------
26. Solubility in Water @ 25 C, ppm(wt)	16.0	90.0	300.0	1080.0

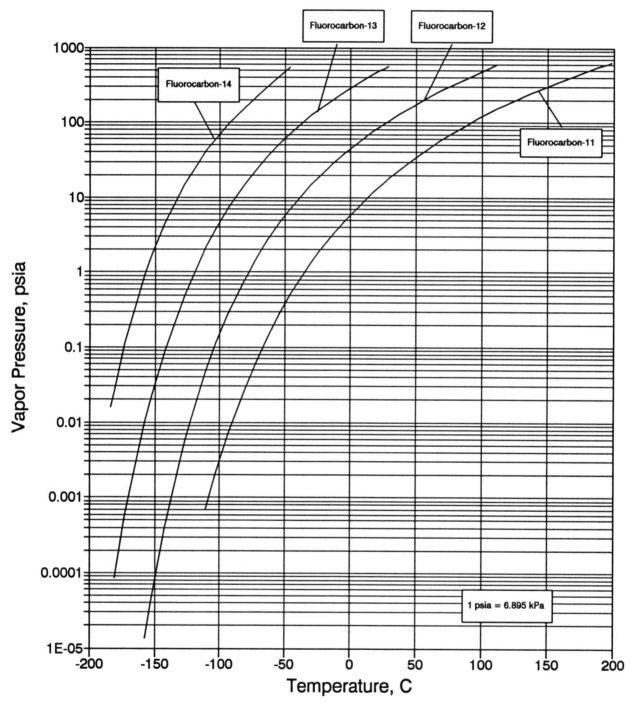

Figure 19-1 Vapor Pressure

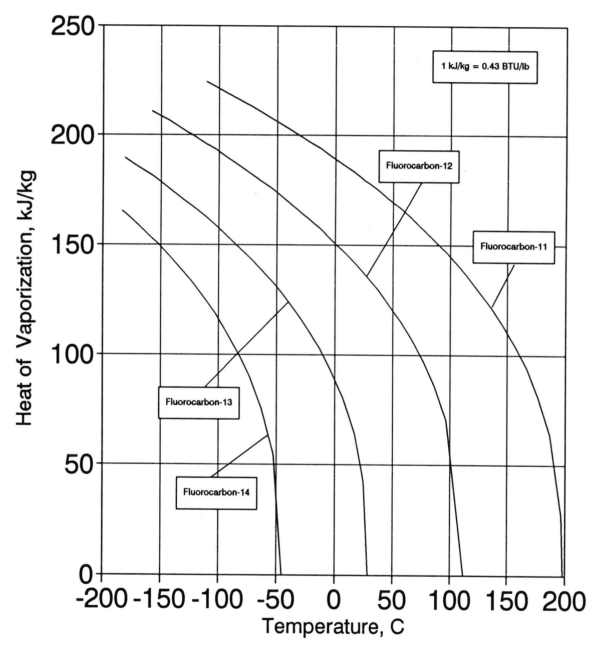

Figure 19-2 Heat of Vaporization

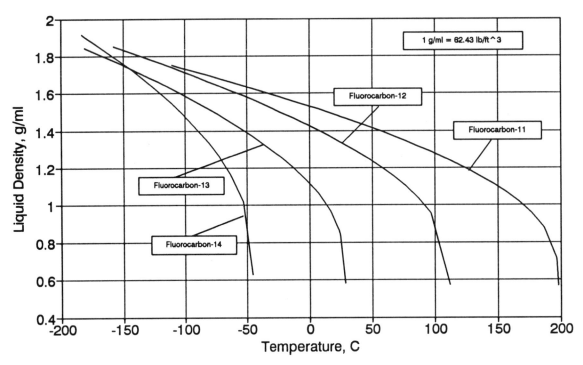

Figure 19-3 Liquid Density

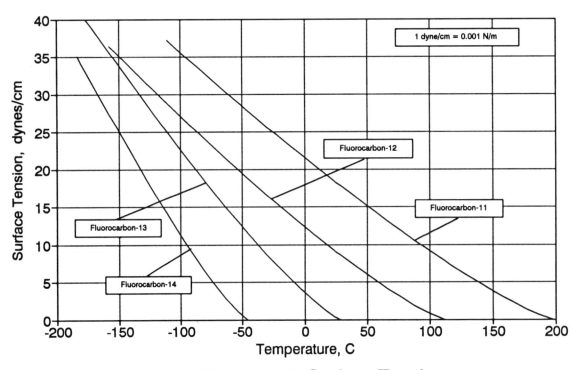

Figure 19-4 Surface Tension

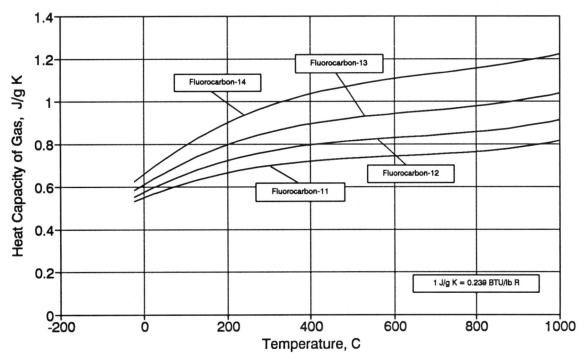

Figure 19-5 Heat Capacity of Gas

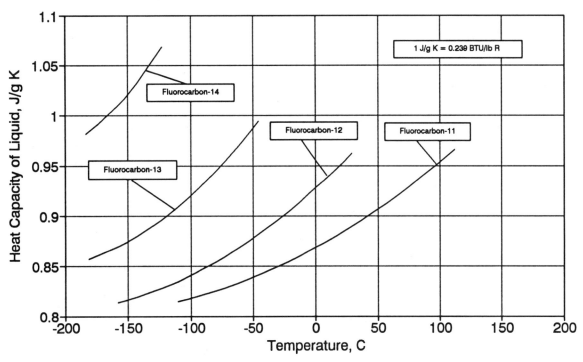

Figure 19-6 Heat Capacity of Liquid

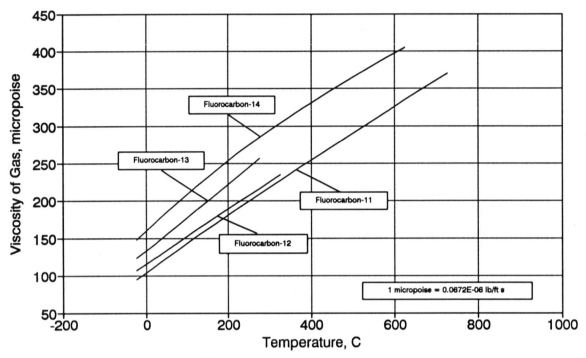

Figure 19-7 Viscosity of Gas

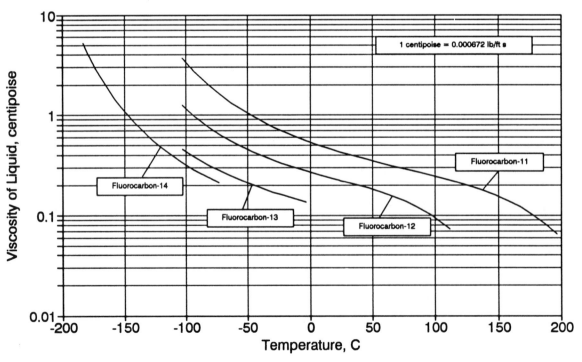

Figure 19-8 Viscosity of Liquid

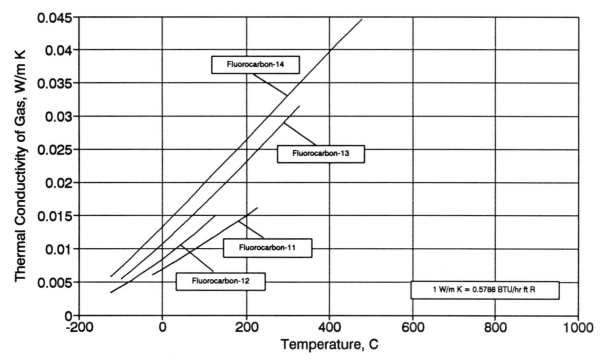

Figure 19-9 Thermal Conductivity of Gas

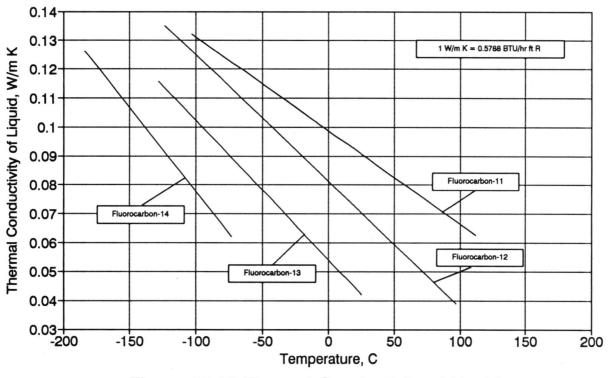

Figure 19-10 Thermal Conductivity of Liquid

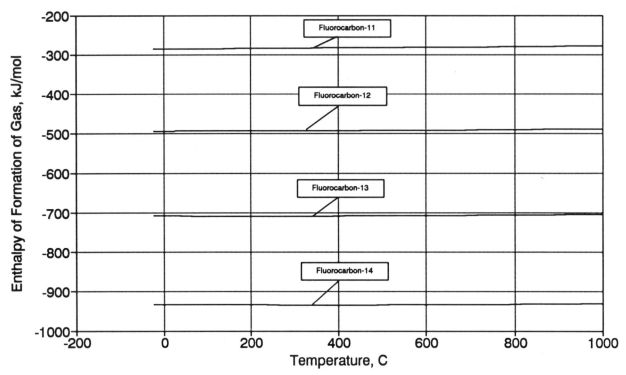

Figure 19-11 Enthalpy of Formation of Gas

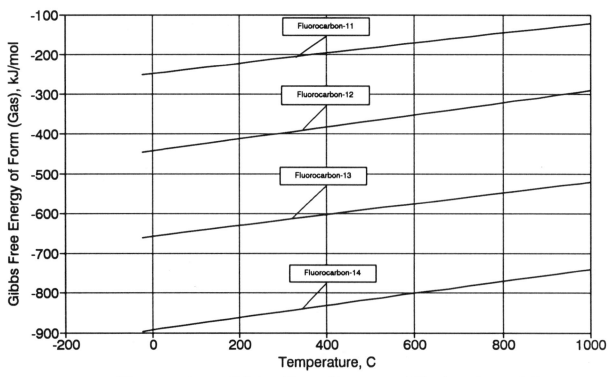

Figure 19-12 Gibbs Free Energy of Formation of Gas

Chapter 20

HALOGENATED HYDROCARBONS

Robert W. Gallant, Carl L. Yaws, Jack. R. Hopper, Xiang Pan and Duane G. Piper, Jr.

PHYSICAL PROPERTIES - Table 20-1

Property data from the literature (1-55,276-289) are given in Table 20-1. The critical constants are from the DIPPR project (5). Additional property data such as acentric factor, enthalpy of formation, lower explosion limit in air and solubility in water are also available. The DIPPR (Design Institute for Physical Property Research) project (5) and recent data compilations by Yaws and co-workers (44-55) were consulted extensively in preparing the tabulation.

VAPOR PRESSURE - Figure 20-1

Results from the DIPPR project (5) were selected for vapor pressure from very low temperatures to the critical point. Correlation of data for vapor pressure as a function of temperature was accomplished using Equation (1-1). Results from this equation (Antoine-type with extended terms) are in favorable agreement with experimental data. Errors are about 1-5% or less in most cases.

HEAT OF VAPORIZATION - Figure 20-2

The data compilation of Yaws and co-workers (44,52) was selected for heat of vaporization for temperatures ranging from melting point to critical point. The Watson equation, Equation (1-2), was used for correlation of the data as a function of temperature. Reliability of results is good with errors of about 1-5% or less.

LIQUID DENSITY - Figure 20-3

Results from the data compilation of Yaws and co-workers (44,54) were selected for liquid density from low temperatures at the melting point to higher temperatures up to the critical point. A modified Rackett equation, Equation (1-3), was used for correlation of the data as a function of temperature. Results from the correlation are in favorable agreement with data. Deviations are less than 1-2% in most cases.

SURFACE TENSION - Figure 20-4

The data compilation of Yaws and co-workers (44,55) was selected for surface tension for temperatures from melting point to critical point. Using data from the literature, correlation for surface tension as a function of temperature over the full liquid range was achieved by the modified Othmer equation, Equation (1-4). Accuracy is good with errors being about 1-10% or less in most cases.

HEAT CAPACITY - Figures 20-5 and 20-6

Results from the data compilation of Yaws and co-workers (44) were selected for heat capacity of ideal gas. Correlation of data was accomplished using a series expansion in temperature, Equation (1-5). Results are in favorable agreement with data. Errors are about 1% or less in most cases.

Results from the DIPPR project (5) were selected for heat capacity of liquid. The coverage applies to temperatures from below the boiling point to temperatures above the boiling point for most of the compounds. Data were correlated with a series expansion in temperature, Equation (1-6). Results are in favorable agreement with data. Errors are about 5% or less using the correlation.

VISCOSITY - Figures 20-7 and 20-8

The DIPPR project (5) was selected for viscosity of gas. Data for gas viscosity as a function of temperature were correlated using Equation (1-7). Results are in favorable agreement with data. Errors are about 1-10% or less in most cases.

The DIPPR project (5) was also selected for viscosity of liquid. Temperatures from below the boiling point to temperatures above the boiling point are covered for most of the compounds. Data for liquid viscosity as a function of temperature were correlated using the de Guzman - Andrade equation with extended terms, Equation (1-8). Correlation results and data are in favorable agreement with errors being about 1-5% or less.

THERMAL CONDUCTIVITY - Figures 20-9 and 20-10

Results from the DIPPR project (5) were selected for thermal conductivity of gas. Data for gas thermal conductivity as a function of temperature were correlated using the Equation (1-9). Reliability of results is good with errors of about 1-10% or less in most cases.

Results from the DIPPR project (5) were selected for thermal conductivity of liquid. The coverage applies to temperatures from below the boiling point to temperatures above the boiling point for most of the compounds. Data for liquid thermal conductivity as a function of temperature were correlated using a series expansion in temperature, Equation (1-10). Results are in favorable agreement with data. Errors are about 1-10% or less in most cases.

ENTHALPY OF FORMATION - Figure 20-11

The data compilation of Yaws and co-workers (44,45) was selected for enthalpy of formation of ideal gas. Data for enthalpy of formation of the ideal gas is a series expansion in temperature, Equation (1-11). Results from the correlation are in favorable agreement with data.

GIBB'S FREE ENERGY OF FORMATION - Figure 20-12

Results from the data compilation of Yaws and co-workers (44,46) were selected for Gibb's free energy of formation of ideal gas. Data for Gibb's free energy of formation of the ideal gas is a series expansion in temperature, Equation (1-12). Correlation results are in favorable agreement with data.

Table 20-1 Physical Properties

1. Name	Fluorocarbon-22 Chlorodifluoro-methane	Fluorocarbon-114 1,2-Dichloro-1,1,2,2-tetra-fluoroethane	Fluorocarbon-21 Dichlorofluoro-methane	Fluorocarbon-113 1,1,2-Trichloro-1,2,2-trifluoro-ethane
2. Formula	$CHClF_2$	$C_2Cl_2F_4$	$CHCl_2F$	$C_2Cl_3F_3$
3. Molecular Weight, g/mol	86.469	170.92	102.92	187.38
4. Critical Temperature, K	369.30	418.85	451.58	487.25
5. Critical Pressure, bar	49.710	32.627	51.838	34.146
6. Critical Volume, ml/mol	166.00	293.68	196.00	325.31
7. Critical Compressibility Factor	0.269	0.275	0.271	0.274
8. Acentric Factor	0.2192	0.252	0.2069	0.2552
9. Melting Point, K	115.73	179.15	138.15	238.15
10. Boiling Point @ 1 atm, K	232.32	276.92	282.05	320.75
11. Heat of Vaporization @ BP, kJ/kg	233.57	136.24	242.32	146.57
12. Density of Liquid @ 25 C, g/ml	1.195	1.418	1.342	1.559
13. Surface Tension @ 25 C, dynes/cm	8.373	20.745	19.631	17.198
14. Heat Capacity of Gas @ 25 C, J/g K	0.646	0.686	0.593	0.679
15. Heat Capacity of Liquid @ 25 C, J/g K	2.086	0.654	1.053	0.908
16. Viscosity of Gas @ 25 C, micropoise	128.97	115.46	115.32	102.84
17. Viscosity of Liquid @ 25 C, centipoise	0.204	0.355	0.321	0.658
18. Thermal Conductivity of Gas @ 25 C, W/m K	0.0110	0.0245	0.0085	0.0076
19. Thermal Conductivity of Liquid @ 25 C, W/m K	0.086	0.063	0.100	0.076
20. Enthalpy of Formation of Gas @ 25 C, kJ/mol	-483.64	-887.55	-284.93	-695.00
21. Gibbs Free Energy of Formation of Gas @ 25 C, kJ/mol	-450.54	-805.51	-252.86	-616.89
22. Flash Point, K	195.00	------	237.00	------
23. Autoignition Temperature, K	905.37	------	825.15	------
24. Lower Explosion Limit in Air, vol %	------	------	------	------
25. Upper Explosion Limit in Air, vol %	26.9	------	54.7	------
26. Solubility in Water @ 25 C, ppm(wt)	2770	137	18800	170

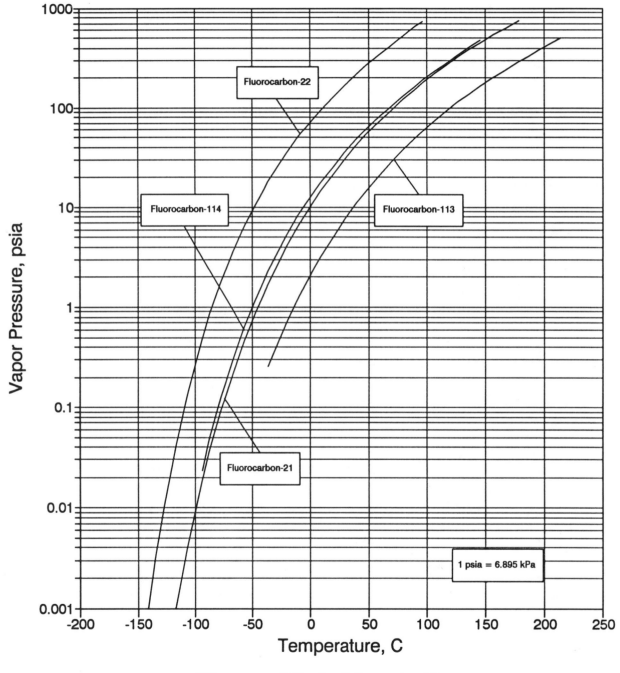

Figure 20-1 Vapor Pressure

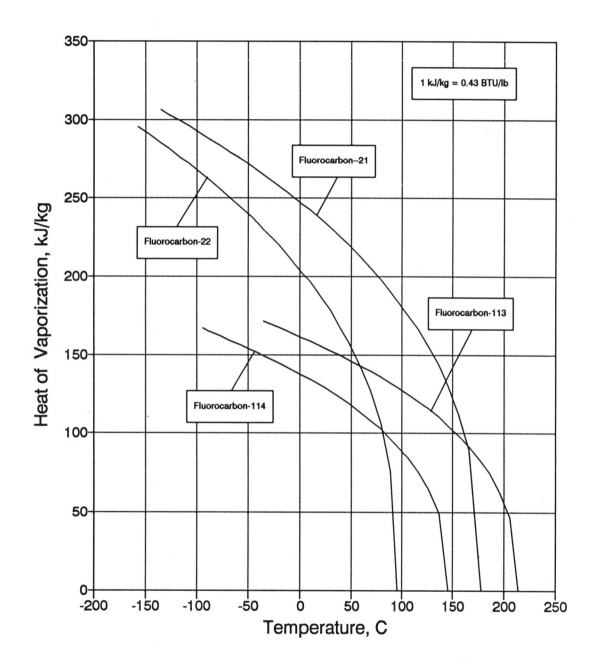

Figure 20-2 Heat of Vaporization

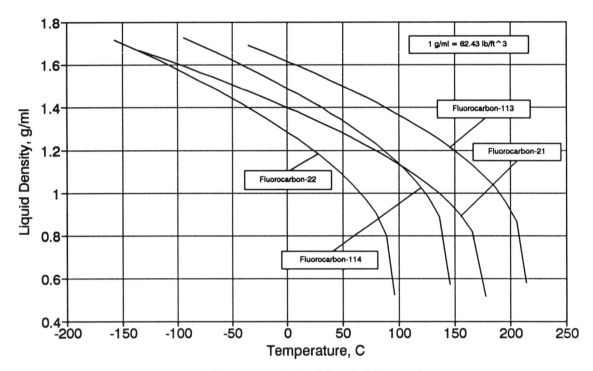

Figure 20-3 Liquid Density

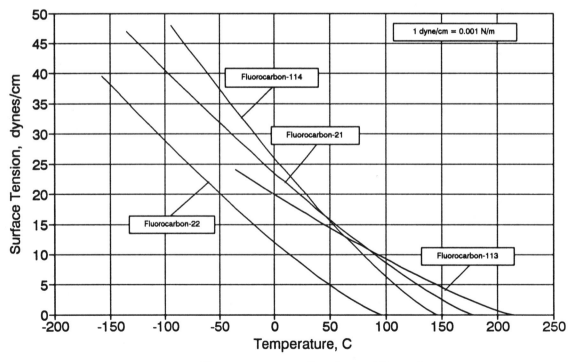

Figure 20-4 Surface Tension

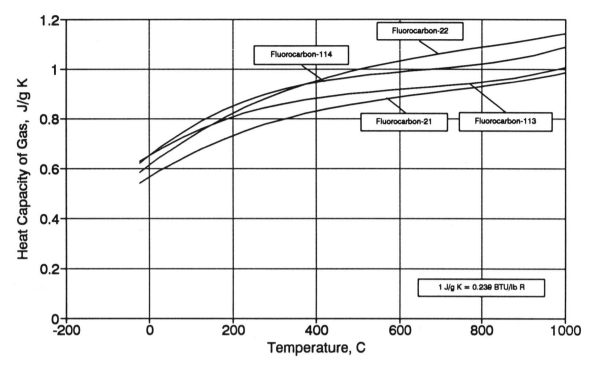

Figure 20-5 Heat Capacity of Gas

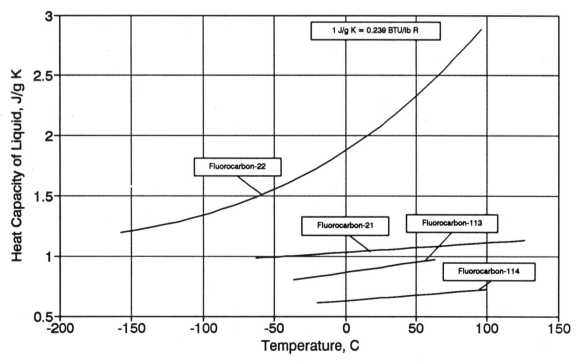

Figure 20-6 Heat Capacity of Liquid

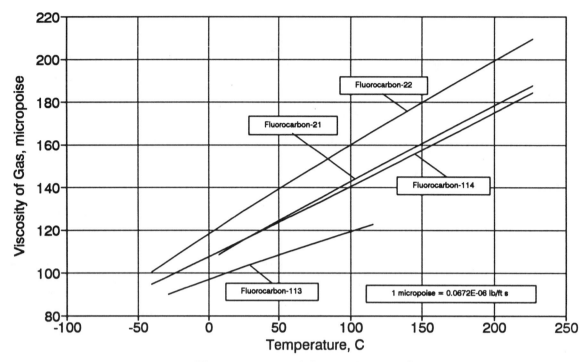

Figure 20-7 Viscosity of Gas

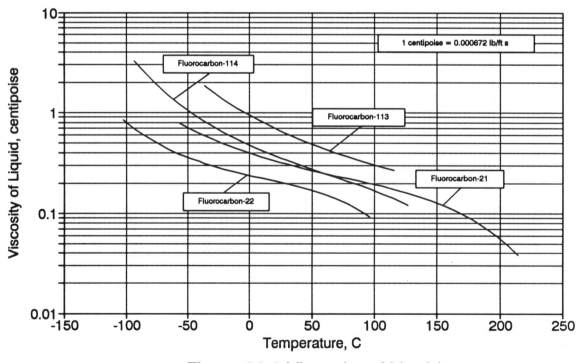

Figure 20-8 Viscosity of Liquid

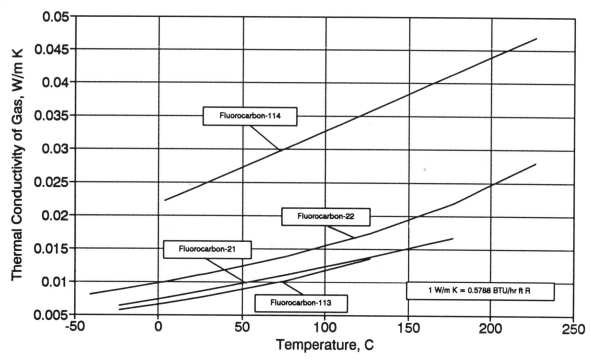

Figure 20-9 Thermal Conductivity of Gas

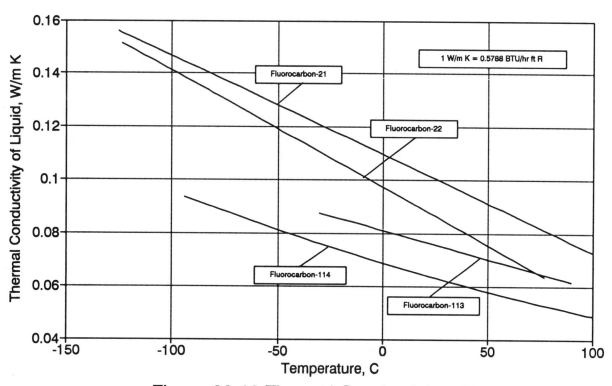

Figure 20-10 Thermal Conductivity of Liquid

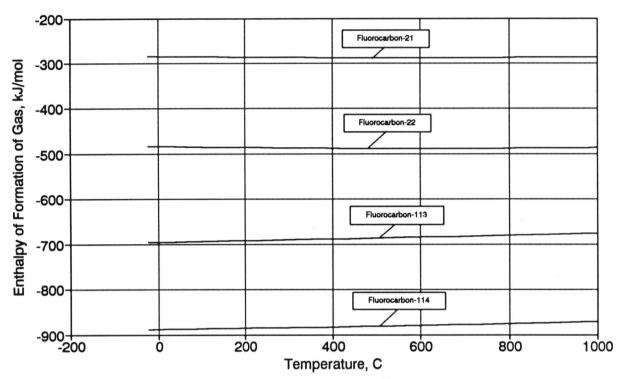

Figure 20-11 Enthalpy of Formation of Gas

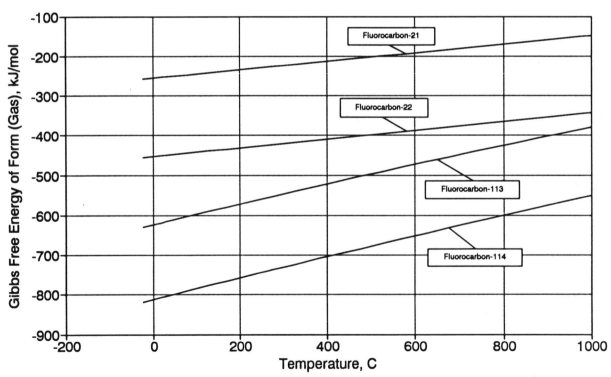

Figure 20-12 Gibbs Free Energy of Formation of Gas

Chapter 21

FLUORINATED HYDROCARBONS

Robert W. Gallant, Carl L. Yaws, Jack R. Hopper, Xiang Pan and Duane G. Piper, Jr.

PHYSICAL PROPERTIES - Table 21-1

Property data from the literature (1-55,276-278,288,290-294) are given in Table 21-1. The critical constants are from the DIPPR project (5). Additional property data such as acentric factor, enthalpy of formation, lower explosion limit in air and solubility in water are also available. The DIPPR (Design Institute for Physical Property Research) project (5) and recent data compilations by Yaws and co-workers (44-55) were consulted extensively in preparing the tabulation.

VAPOR PRESSURE - Figure 21-1

Results from the DIPPR project (5) were selected for vapor pressure from very low temperatures to the critical point. Correlation of data for vapor pressure as a function of temperature was accomplished using Equation (1-1). Results from this equation (Antoine-type with extended terms) are in favorable agreement with experimental data. Errors are about 1-5% or less in most cases.

HEAT OF VAPORIZATION - Figure 21-2

The data compilation of Yaws and co-workers (44,52) was selected for heat of vaporization for temperatures ranging from melting point to critical point. The Watson equation, Equation (1-2), was used for correlation of the data as a function of temperature. Reliability of results is good with errors of about 1-5% or less.

LIQUID DENSITY - Figure 21-3

Results from the DIPPR project (5) were selected for liquid density from low temperatures at the melting point to higher temperatures up to the critical point. A modified Rackett equation, Equation (1-3a), was used for correlation of the data as a function of temperature. Results from the correlation are in favorable agreement with data. Deviations are less than 1-2% in most cases.

SURFACE TENSION - Figure 21-4

The data compilation of Yaws and co-workers (44,55) was selected for surface tension for temperatures from melting point to critical point. Using data from the literature, correlation for surface tension as a function of temperature over the full liquid range was achieved by the modified Othmer equation, Equation (1-4). Accuracy is good with errors being about 1-10% or less in most cases.

HEAT CAPACITY - Figures 21-5 and 21-6

Results from the DIPPR project (5) were selected for heat capacity of ideal gas. Correlation of data was accomplished using Equation (1-5a). Results are in favorable agreement with data. Errors are about 1% or less in most cases.

Results from the DIPPR project (5) were selected for heat capacity of liquid. The coverage applies to temperatures from below the boiling point to temperatures above the boiling point for most of the compounds. Data were correlated with a series expansion in temperature, Equation (1-6). Results are in favorable agreement with data. Errors are about 5% or less using the correlation.

VISCOSITY - Figures 21-7 and 21-8

The DIPPR project (5) was selected for viscosity of gas. Data for gas viscosity as a function of temperature were correlated using Equation (1-7). Results are in favorable agreement with data. Errors are about 1-10% or less in most cases.

The DIPPR project (5) was also selected for viscosity of liquid. Temperatures from below the boiling point to temperatures above the boiling point are covered for most of the compounds. Data for liquid viscosity as a function of temperature were correlated using the de Guzman - Andrade equation with extended terms, Equation (1-8). Correlation results and data are in favorable agreement with errors being about 1-5% or less.

THERMAL CONDUCTIVITY - Figures 21-9 and 21-10

Results from the DIPPR project (5) were selected for thermal conductivity of gas. Data for gas thermal conductivity as a function of temperature were correlated using the Equation (1-9). Reliability of results is good with errors of about 1-10% or less in most cases except for hexafluoroethane (possible 25% error).

Results from the DIPPR project (5) were selected for thermal conductivity of liquid. The coverage applies to temperatures from below the boiling point to temperatures above the boiling point for most of the compounds. Data for liquid thermal conductivity as a function of temperature were correlated using a series expansion in temperature, Equation (1-10). Results are in favorable agreement with data. Errors are about 1-10% or less in most cases.

ENTHALPY OF FORMATION - Figure 21-11

The data compilation of Yaws and co-workers (44,45) was selected for enthalpy of formation of ideal gas. Data for enthalpy of formation of the ideal gas is a series expansion in temperature, Equation (1-11). Results from the correlation are in favorable agreement with data.

GIBB'S FREE ENERGY OF FORMATION - Figure 21-12

Results from the data compilation of Yaws and co-workers (44,46) were selected for Gibb's free energy of formation of ideal gas. Data for Gibb's free energy of formation of the ideal gas is a series expansion in temperature, Equation (1-12). Correlation results are in favorable agreement with data.

Table 21-1 Physical Properties

1. Name	Trifluoro-methane Fluorocarbon-23	Hexafluoro-ethane Fluorocarbon-116	Vinyl Fluoride	Vinylidene Fluoride
2. Formula	CHF_3	C_2F_6	C_2H_3F	$C_2H_2F_2$
3. Molecular Weight, g/mol	70.014	138.010	46.044	64.035
4. Critical Temperature, K	298.89	292.80	327.80	302.80
5. Critical Pressure, bar	48.362	29.789	52.385	44.583
6. Critical Volume, ml/mol	133.30	224.00	144.00	154.00
7. Critical Compressibility Factor	0.259	0.274	0.277	0.273
8. Acentric Factor	0.2672	0.2452	0.2151	0.1390
9. Melting Point, K	117.97	172.45	112.65	129.15
10. Boiling Point @ 1 atm, K	190.99	194.95	200.95	187.50
11. Heat of Vaporization @ BP, kJ/kg	241.40	96.38	397.10	337.90
12. Density of Liquid @ 25 C, g/ml	0.667	------	0.620	0.594
13. Surface Tension @ 25 C, dynes/cm	0.052	------	3.056	0.446
14. Heat Capacity of Gas @ 25 C, J/g K	0.732	0.778	1.096	0.905
15. Heat Capacity of Liquid @ 25 C, J/g K	------	------	------	------
16. Viscosity of Gas @ 25 C, micropoise	143.94	148.07	115.39	131.04
17. Viscosity of Liquid @ 25 C, centipoise	0.148	------	------	------
18. Thermal Conductivity of Gas @ 25 C, W/m K	0.0122	0.0348	0.0144	0.0153
19. Thermal Conductivity of Liquid @ 25 C, W/m K	------	------	0.088	------
20. Enthalpy of Formation of Gas @ 25 C, kJ/mol	-697.55	-1342.20	-138.94	-336.86
21. Gibbs Free Energy of Formation of Gas @ 25 C, kJ/mol	-663.17	-1257.40	-125.17	-313.15
22. Flash Point, K	161.00	------	------	------
23. Autoignition Temperature, K	------	------	658.15	913.00
24. Lower Explosion Limit in Air, vol %	------	------	2.6	21.3
25. Upper Explosion Limit in Air, vol %	35.3	------	21.7	28.39
26. Solubility in Water @ 25 C, ppm(wt)	900	7.879	------	164.9

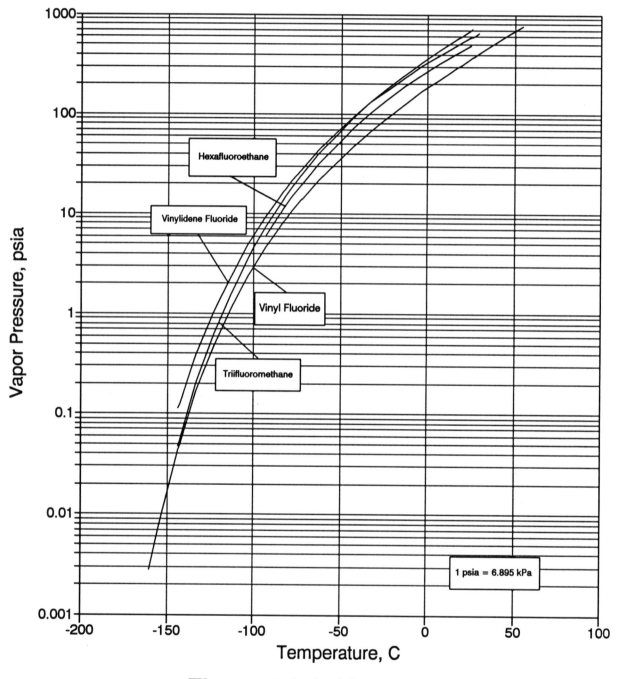

Figure 21-1 Vapor Pressure

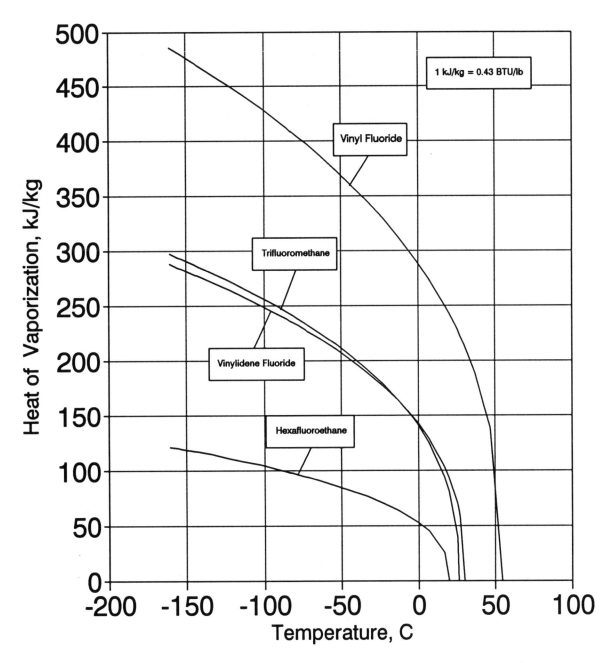

Figure 21-2 Heat of Vaporization

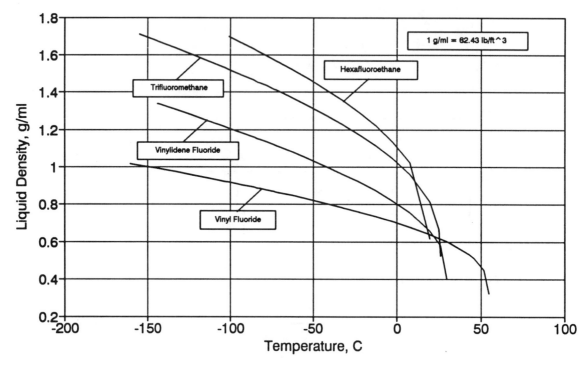

Figure 21-3 Liquid Density

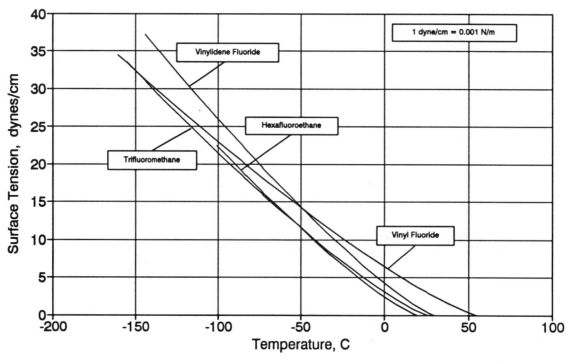

Figure 21-4 Surface Tension

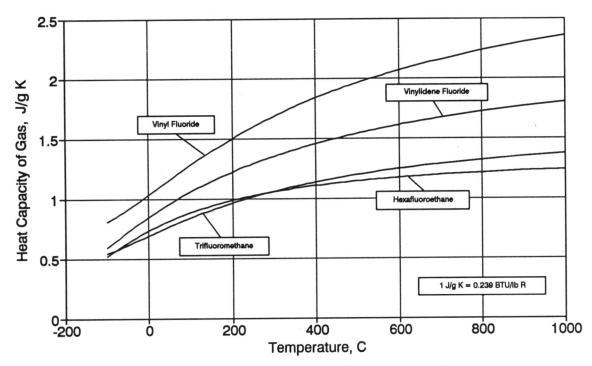

Figure 21-5 Heat Capacity of Gas

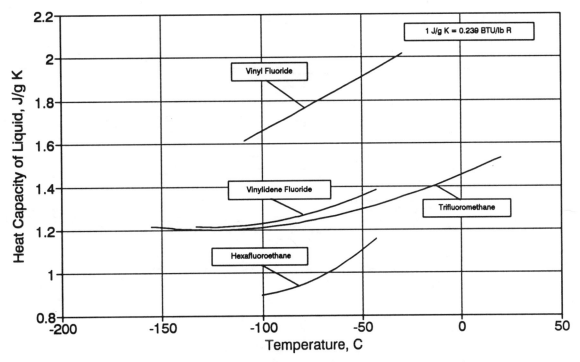

Figure 21-6 Heat Capacity of Liquid

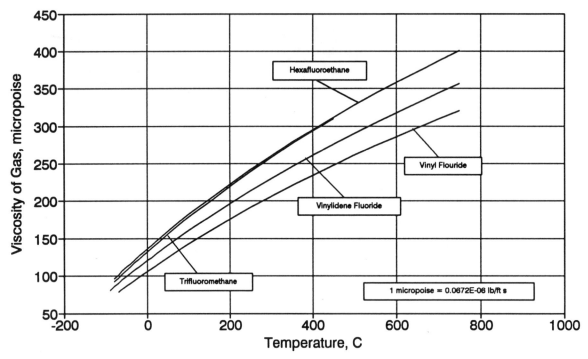

Figure 21-7 Viscosity of Gas

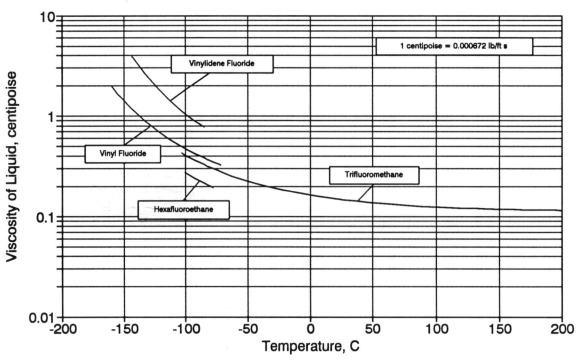

Figure 21-8 Viscosity of Liquid

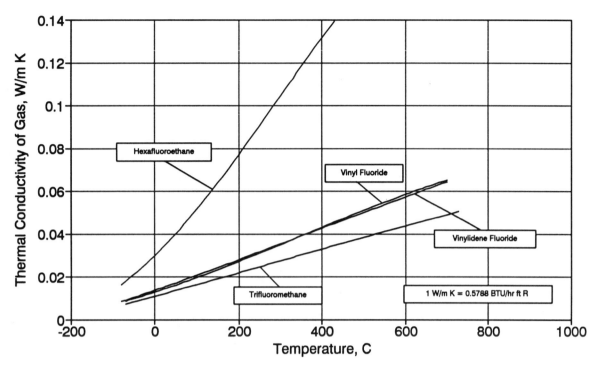

Figure 21-9 Thermal Conductivity of Gas

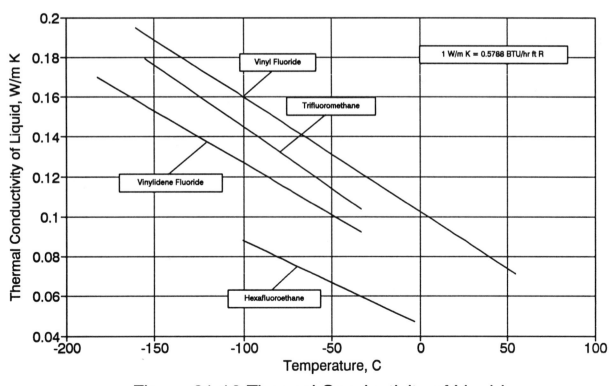

Figure 21-10 Thermal Conductivity of Liquid

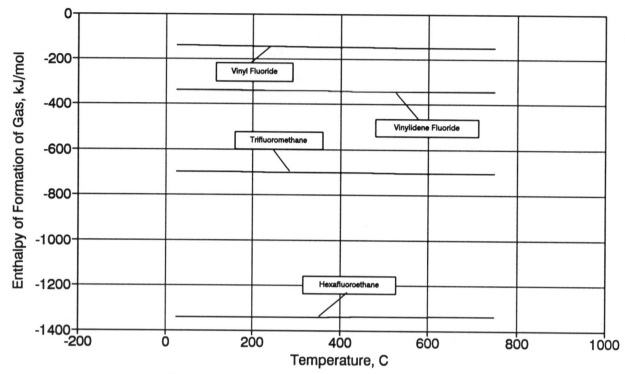

Figure 21-11 Enthalpy of Formation of Gas

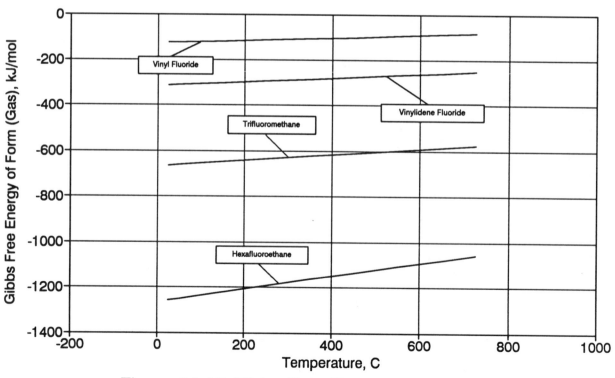

Figure 21-12 Gibbs Free Energy of Formation of Gas

Chapter 22

BROMINATED HYDROCARBONS

Robert W. Gallant, Carl L. Yaws, Jack R. Hopper, Xiang Pan and Duane G. Piper, Jr.

PHYSICAL PROPERTIES - Table 22-1

Property data from the literature (1-55,111,112,124,125,139,241,278,295-311) are given in Table 21-1. The critical constants are from the DIPPR project (5). Additional property data such as acentric factor, enthalpy of formation, lower explosion limit in air and solubility in water are also available. The DIPPR (Design Institute for Physical Property Research) project (5) and recent data compilations by Yaws and co-workers (44-55) were consulted extensively in preparing the tabulation.

VAPOR PRESSURE - Figure 22-1

Results from the DIPPR project (5) were selected for vapor pressure from very low temperatures to the critical point. Correlation of data for vapor pressure as a function of temperature was accomplished using Equation (1-1). Results from this equation (Antoine-type with extended terms) are in favorable agreement with experimental data. Errors are about 1-5% or less in most cases.

HEAT OF VAPORIZATION - Figure 22-2

The data compilation of Yaws and co-workers (44,52) was selected for heat of vaporization for temperatures ranging from melting point to critical point. The Watson equation, Equation (1-2), was used for correlation of the data as a function of temperature. Reliability of results is good with errors of about 1-5% or less.

LIQUID DENSITY - Figure 22-3

Results from the DIPPR project (5) were selected for liquid density from low temperatures at the melting point to higher temperatures up to the critical point. A modified Rackett equation, Equation (1-3a), was used for correlation of the data as a function of temperature. Results from the correlation are in favorable agreement with data. Deviations are less than 1-2% in most cases.

SURFACE TENSION - Figure 22-4

The data compilation of Yaws and co-workers (44,55) was selected for surface tension for temperatures from melting point to critical point. Using data from the literature, correlation for surface tension as a function of temperature over the full liquid range was achieved by the modified Othmer equation, Equation (1-4). Accuracy is good with errors being about 1-10% or less in most cases.

HEAT CAPACITY - Figures 22-5 and 22-6

Results from the DIPPR project (5) were selected for heat capacity of ideal gas. Correlation of data was accomplished using Equation (1-5a). Results are in favorable agreement with data. Errors are about 1% or less in most cases.

Results from the DIPPR project (5) were selected for heat capacity of liquid. The coverage applies to temperatures from below the boiling point to temperatures above the boiling point for most of the compounds. Data were correlated with a series expansion in temperature, Equation (1-6). Results are in favorable agreement with data. Errors are about 5% or less using the correlation.

VISCOSITY - Figures 22-7 and 22-8

The DIPPR project (5) was selected for viscosity of gas. Data for gas viscosity as a function of temperature were correlated using Equation (1-7). Results are in favorable agreement with data. Errors are about 1-10% or less in most cases.

The DIPPR project (5) was also selected for viscosity of liquid. Temperatures from below the boiling point to temperatures above the boiling point are covered for most of the compounds. Data for liquid viscosity as a function of temperature were correlated using the de Guzman - Andrade equation with extended terms, Equation (1-8). Correlation results and data are in favorable agreement with errors being about 1-10% or less.

THERMAL CONDUCTIVITY - Figures 22-9 and 22-10

Results from the DIPPR project (5) were selected for thermal conductivity of gas. Data for gas thermal conductivity as a function of temperature were correlated using the Equation (1-9). Reliability of results is good with errors of about 1-10% or less in most cases except for ethylene dibromide (possible 25% error).

Results from the DIPPR project (5) were selected for thermal conductivity of liquid. The coverage applies to temperatures from below the boiling point to temperatures above the boiling point for most of the compounds. Data for liquid thermal conductivity as a function of temperature were correlated using a series expansion in temperature, Equation (1-10). Results are in favorable agreement with data. Errors are about 1-10% or less in most cases.

ENTHALPY OF FORMATION - Figure 22-11

The data compilation of Yaws and co-workers (44,45) was selected for enthalpy of formation of ideal gas except for bromotrifluoromethane (3). Data for enthalpy of formation of the ideal gas is a series expansion in temperature, Equation (1-11). Results from the correlation are in favorable agreement with data.

GIBB'S FREE ENERGY OF FORMATION - Figure 22-12

Results from the data compilation of Yaws and co-workers (44,46) were selected for Gibb's free energy of formation of ideal gas except for bromotrifluoromethane (3). Data for Gibb's free energy of formation of the ideal gas is a series expansion in temperature, Equation (1-12). Correlation results are in favorable agreement with data.

Table 22-1 Physical Properties

	Bromotrifluoro-methane	Methyl Bromide	Ethyl Bromide	Ethylene Dibromide
1. Name	Bromotrifluoro-methane	Methyl Bromide	Ethyl Bromide	Ethylene Dibromide
2. Formula	$CBrF_3$	CH_3Br	C_2H_5Br	$C_2H_4Br_2$
3. Molecular Weight, g/mol	148.910	94.939	108.970	187.860
4. Critical Temperature, K	340.15	467.00	503.80	650.15
5. Critical Pressure, bar	39.719	52.300	62.315	54.769
6. Critical Volume, ml/mol	200.00	156.00	214.92	261.57
7. Critical Compressibility Factor	0.281	0.210	0.320	0.265
8. Acentric Factor	0.1727	0.1922	0.2533	0.2067
9. Melting Point, K	105.15	179.55	154.55	282.94
10. Boiling Point @ 1 atm, K	215.26	276.71	311.50	404.51
11. Heat of Vaporization @ BP, kJ/kg	117.06	214.28	244.36	183.08
12. Density of Liquid @ 25 C, g/ml	1.536	1.662	1.450	2.169
13. Surface Tension @ 25 C, dynes/cm	4.168	18.345	23.507	38.158
14. Heat Capacity of Gas @ 25 C, J/g K	0.468	0.446	0.594	0.454
15. Heat Capacity of Liquid @ 25 C, J/g K	0.869	------	0.916	0.725
16. Viscosity of Gas @ 25 C, micropoise	156.80	132.60	110.22	106.73
17. Viscosity of Liquid @ 25 C, centipoise	0.159	0.310	0.374	1.600
18. Thermal Conductivity of Gas @ 25 C, W/m K	0.0098	0.0075	0.0081	0.0057
19. Thermal Conductivity of Liquid @ 25 C, W/m K	0.0407	0.1022	0.1018	0.1007
20. Enthalpy of Formation of Gas @ 25 C, kJ/mol	-649.80	-37.66	-64.02	-38.91
21. Gibbs Free Energy of Formation of Gas @ 25 C, kJ/mol	-622.90	-28.16	-26.32	-10.59
22. Flash Point, K	------	229.00	240.00	------
23. Autoignition Temperature, K	------	810.37	784.26	------
24. Lower Explosion Limit in Air, vol %	------	10.0	6.7	------
25. Upper Explosion Limit in Air, vol %	------	16.0	11.3	------
26. Solubility in Water @ 25 C, ppm(wt)	------	13410	9000	4170

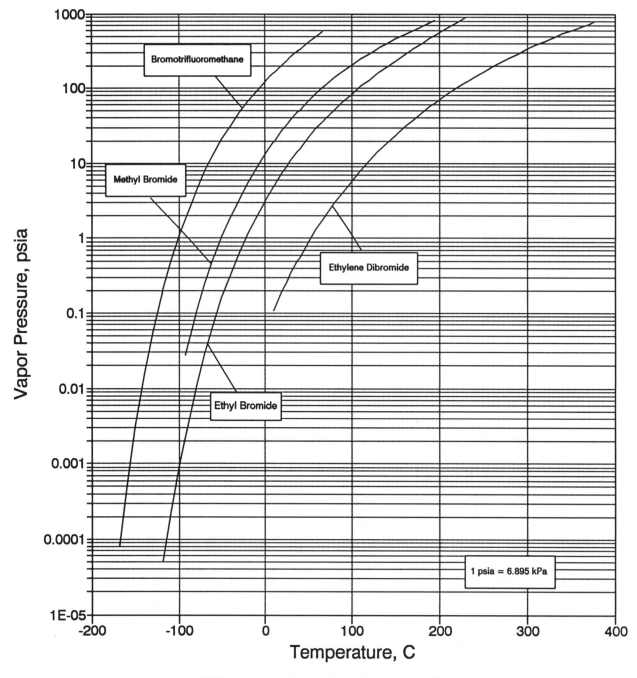

Figure 22-1 Vapor Pressure

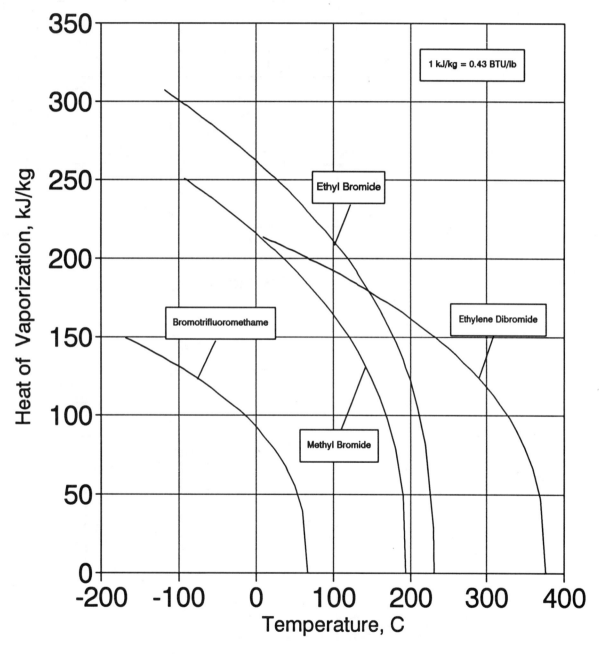

Figure 22-2 Heat of Vaporization

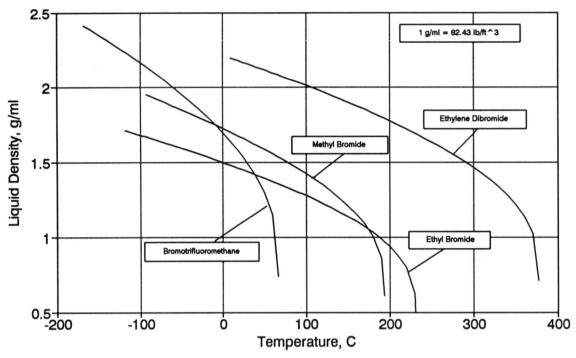

Figure 22-3 Liquid Density

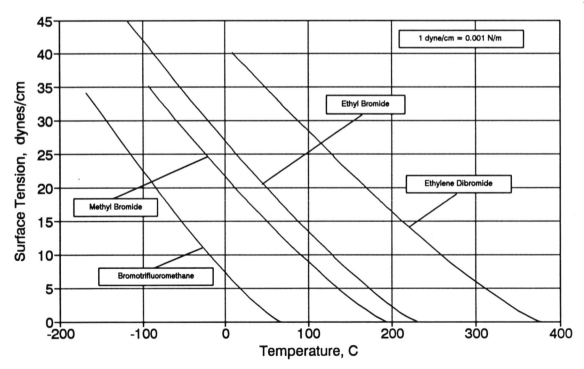

Figure 22-4 Surface Tension

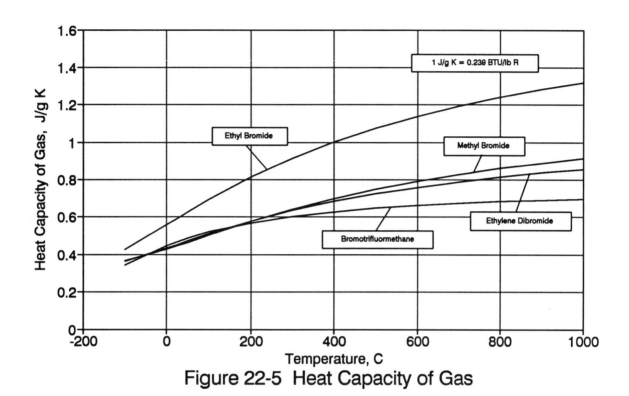

Figure 22-5 Heat Capacity of Gas

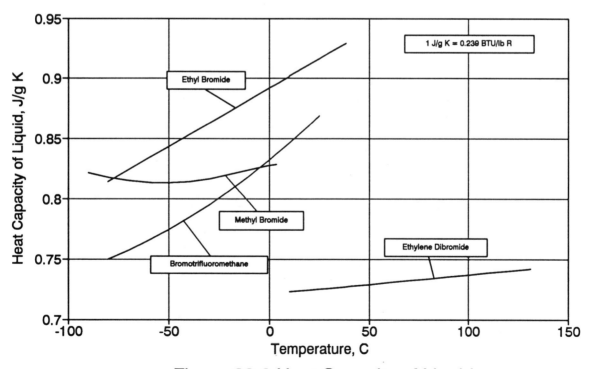

Figure 22-6 Heat Capacity of Liquid

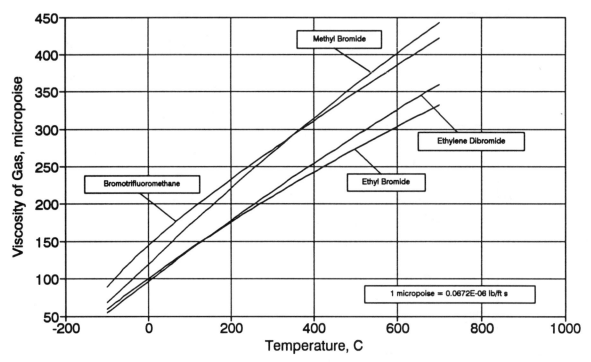

Figure 22-7 Viscosity of Gas

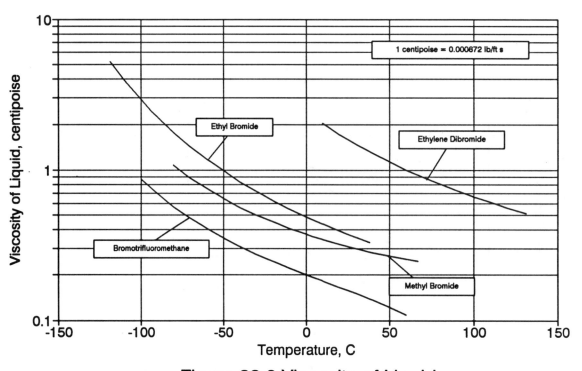

Figure 22-8 Viscosity of Liquid

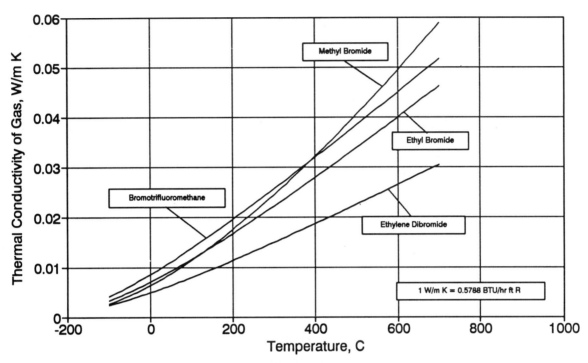

Figure 22-9 Thermal Conductivity of Gas

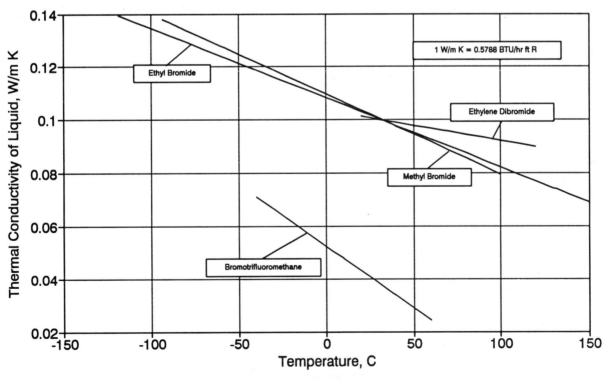

Figure 22-10 Thermal Conductivity of Liquid

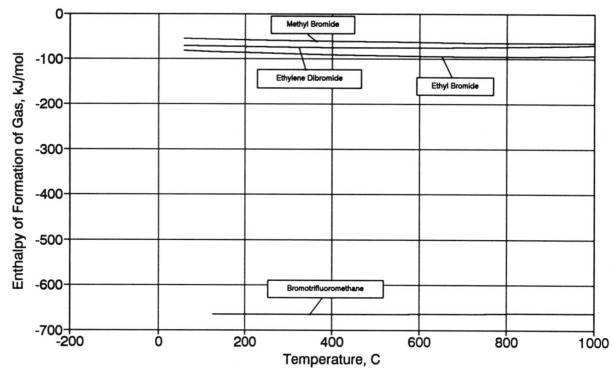

Figure 22-11 Enthalpy of Formation of Gas

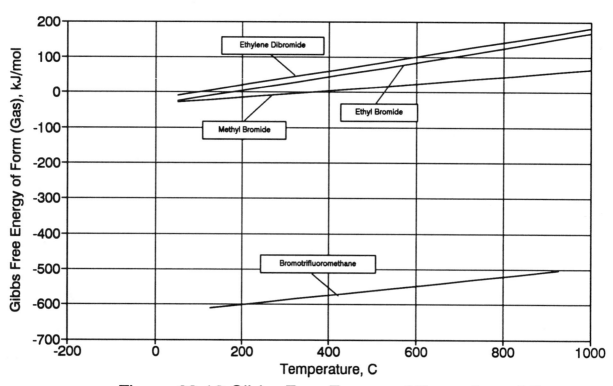

Figure 22-12 Gibbs Free Energy of Formation of Gas

REFERENCES

1. API Research Project No. 44, SELECTED VALUES OF PHYSICAL AND THERMODYNAMIC PROPERTIES OF HYDROCARBONS AND RELATED COMPOUNDS, Carnegie Press, Carnegie Institute of Technology, Pittsburgh, PA (1953).
2. SELECTED VALUES OF PROPERTIES OF HYDROCARBONS AND RELATED COMPOUNDS, Thermodynamics Research Center, TAMU, College Station, TX (1977, 1984).
3. SELECTED VALUES OF PROPERTIES OF CHEMICAL COMPOUNDS, Thermodynamics Research Center, TAMU, College Station, TX (1977, 1987).
4. TECHNICAL DATA BOOK - PETROLEUM REFINING, Vol. I and II, American Petroleum Institute, Washington, DC (1972, 1977, 1982).
5. Daubert, T. E. and R. P. Danner, DATA COMPILATION OF PROPERTIES OF PURE COMPOUNDS, Parts 1 and 2, DIPPR Project, AIChE, New York, NY (1985-present).
6. Ambrose, D., VAPOUR-LIQUID CRITICAL PROPERTIES, National Physical Laboratory, Teddington, England, NPL Report Chem 107 (Feb., 1980).
7. Simmrock, K. H., R. Janowsky and A. Ohnsorge, CRITICAL DATA OF PURE SUBSTANCES, Vol. II, Parts 1 and 2, Dechema Chemistry Data Series, 6000 Frankfurt/Main, Germany (1986).
8. INTERNATIONAL CRITICAL TABLES, McGraw-Hill, New York, NY (1926).
9. Maxwell, J. B., DATA BOOK ON HYDROCARBONS, D. Van Nostrand, Princeton. NJ (1958).
10. Egloff, G., PHYSICAL CONSTANTS OF HYDROCARBONS, Vols. 1-6, Reinhold Publishing Corp., New York, NY (1939-1947).
11. Din, F., THERMODYNAMIC FUNCTIONS OF GASES, Vols. 1-3, Butterworth Scientific Publications, London, England (1956-1962).
12. Edmister, W. C., APPLIED HYDROCARBON THERMODYNAMICS, Vols. 1 and 2, Vol 2, 2nd ed., Gulf Publishing Co., Houston, TX (1961, 1974, 1984).
13. Braker, W. and A. L. Mossman, MATHESON GAS DATA BOOK, 6th ed., Matheson Gas Products, Secaucaus, NJ (1980).
14. CRC HANDBOOK OF CHEMISTRY AND PHYSICS, 66th - 71th ed., CRC Press, Inc., Boca Raton, FL (1985-1990).
15. LANGE'S HANDBOOK OF CHEMISTRY, 13th ed., McGraw-Hill, New York, NY (1985).
16. PERRY'S CHEMICAL ENGINEERING HANDBOOK, 6th ed., McGraw-Hill, New York, NY (1984).
17. Kaye, G. W. C. and T. H. Laby, TABLES OF PHYSICAL AND CHEMICAL CONSTANTS, Longman Group Limited, London, England (1973).
18. Raznjevic, Kuzman, HANDBOOK OF THERMODYNAMIC TABLES AND CHARTS, Hemisphere Publishing Corporation, New York, NY (1976).
19. Driesbach, R. R., PHYSICAL PROPERTIES OF CHEMICAL COMPOUNDS, Vol. I (No. 15), Vol. II (No. 22), Vol. III (No. 29), Advances in Chemistry Series, American Chemical Society, Washington, DC (1955,1959,1961).
20. Vargaftik, N. B., TABLES ON THE THERMOPHYSICAL PROPERTIES OF LIQUIDS AND GASES, 2nd ed., English translation, Hemisphere Publishing Corporation, New York, NY (1975, 1983).
21. Timmermans, J., PHYSICO-CHEMICAL CONSTANTS OF PURE ORGANIC COMPOUNDS, Vol. 1 and 2, Elsevier, New York, NY (1950,1965).
22. Lyman, W. J., W. F. Reehl and D. H. Rosenblatt, HANDBOOK OF CHEMICAL PROPERTY ESTIMATION METHODS, McGraw-Hill, New York, NY (1982).
23. Bretsznajder, S., PREDICTION OF TRANSPORT AND OTHER PHYSICAL PROPERTIES OF FLUIDS, International Series of Monographs in Chemical Engineering, Vol. 2, Pergamon Press, Oxford, England (1971).
24. Reid, R. C. and T. K. Sherwood, THE PROPERTIES OF GASES AND LIQUIDS, 3rd ed., McGraw-Hill, New York, NY (1977).

25. Reid, R. C., J. M. Prausnitz and B. E. Poling, THE PROPERTIES OF GASES AND LIQUIDS, 4th ed., McGraw-Hill, New York, NY (1987).
26. Kirk, R. E. and D. F. Othmer, editors, ENCYCLOPEDIA OF CHEMICAL TECHNOLOGY, 3rd ed., Vols. 1-24, John Wiley and Sons, Inc., New York, NY (1978-1984).
27. Sax, N. I. and R. J. Lewis, Jr., HAWLEY'S CONDENSED CHEMICAL DICTIONARY, 11th ed., Van Nostrand Reinhold Co., New York, NY (1987).
28. Bondi, A., PHYSICAL PROPERTIES OF MOLECULAR CRYSTALS, LIQUIDS AND GLASSES, John Wiley, New York, NY (1968).
29. Beaton, C. F. and G. F. Hewitt, PHYSICAL PROPERTY DATA FOR THE DESIGN ENGINEER, Hemisphere Publishing Corporation, New York, NY (1989).
30. Gallant, R. W., PHYSICAL PROPERTIES OF HYDROCARBONS, Vols. 1 and 2, Vol 2, 2nd ed., Gulf Publishing Co., Houston, TX (1968, 1970, 1984).
31. Yaws, C. L., PHYSICAL PROPERTIES, McGraw-Hill, New York, NY (1977).
32. Zwolinski, B. J. and R. C. Wilhoit, VAPOR PRESSURES AND HEATS OF VAPORIZATION OF HYDROCARBONS AND RELATED COMPOUNDS, Thermodynamic Research Center, TAMU, College Station, TX (1971).
33. Boublick, T., V. Fried and E. Hala, THE VAPOUR PRESSURES OF PURE SUBSTANCES, 1st ed., 2nd ed., Elsevier, New York, NY (1975, 1984).
34. Ohe, S., COMPUTER AIDED DATA BOOK OF VAPOR PRESSURE, Data Book Publishing Company, Tokyo, Japan (1976).
35. Tryon, G. H., ed., FIRE PROTECTION HANDBOOK, 12th ed., National Fire Protection Association, Boston, Mass. (1962).
36. Jasper, J.J., J. Phys. Chem. Ref. Data, 1 (No.4), 841 (1972).
37. JANAF THERMOCHEMICAL TABLES, 2nd edition, NSRDS-NBS 37, U. S. Government Printing Office, Washington DC (1971).
38. Stull, D. R., E. F. Westrum, Jr., and G. C. Sinke, THE CHEMICAL THERMODYNAMICS OF ORGANIC COMPOUNDS, John Wiley and Sons, New York, NY (1969).
39. THERMOPHYSICAL PROPERTIES OF MATTER, 1st and 2nd eds., IFI/Plenum, New York, NY (1970-1976).
40. Ho, C. Y., P. E. Liley, T. Makita and Y. Tanaka, PROPERTIES OF INORGANIC AND ORGANIC FLUIDS, Hemisphere Publishing Corporation, New York, NY (1988).
41. Golubev, I. F., VISCOSITY OF GASES AND GAS MIXTURES, translated from Russian, US Dept. of Commerce, Springfield, VA (1970).
42. Stephan, K. and Lucas K., VISCOSITY OF DENSE FLUIDS, Plenum Press, New York, NY (1979).
43. Viswanath, D. S. and G. Natarajan, DATA BOOK ON THE VISCOSITY OF LIQUIDS, Hemisphere Publishing Corporation, New York, NY (1989).
44. Yaws, C. L., THERMODYNAMIC AND PHYSICAL PROPERTY DATA, Gulf Publishing Co., Houston, TX (1992).
45. Yaws, C. L. and P. Y. Chiang, "Enthalpy of Formation for 700 Major Organic Compounds," Chem. Eng., 95, 81 (Sept. 26, 1988).
46. Yaws, C. L. and P. Y. Chiang, "Find Favorable Reactions Faster," Hydrocarbon Processing, 67, 81 (Nov., 1988).
47. Yaws, C. L., D. Chen, H.-C. Yang, L. Tan and D. Nico, "Critical Properties of Chemicals", Hydrocarbon Processing, 68, 61 (July, 1989).
48. Yaws, C. L. and H.-C. Yang, "To Estimate Vapor Pressure Easily", Hydrocarbon Processing, 68, 65 (Oct.,1989).
49. Yaws, C. L., H.-C. Yang, Jack R. Hopper and Keith C. Hansen, "Hydrocarbons: Water Solubility Data", Chem. Eng., 97, 177 (April,1990).
50. Yaws, C. L., H.-C. Yang, Jack R. Hopper and Keith C. Hansen, "LETTERS: Hydrocarbons - Water

Solubility Data", Chem. Eng., 97, 8 (May, 1990).

51. Yaws, C. L., H.-C. Yang, Jack R. Hopper and Keith C. Hansen, "Organic Chemicals: Water Solubility Data", Chem. Eng., 97, 115 (July, 1990).
52. Yaws, C. L., H.-C. Yang and William C. Cawley, "Predict Enthalpy of Vaporization", Hydrocarbon Processing, 69, 87 (June, 1990).
53. Yaws, C. L. and H.-C. Yang, "Water Solubility Data for Organic Compounds", Pollution Engineering, 22, 70 (Oct., 1990).
54. Yaws, C. L., H.-C. Yang, Jack R. Hopper and William C. Cawley, "An Equation for Liquid Density", Hydrocarbon Processing, 70, 103 (January, 1991).
55. Yaws, C. L., H.-C. Yang and Xiang Pan, "633 Organic Chemicals: Surface Tension Data", Chem. Eng., 98, 140 (March, 1991).
56. Goodwin, R. D., H. M. Roder and G. C. Straty, National Bureau of Standards Technical Note 684, Cryogenics Division, Boulder, CO (1976).
57. Perry, R. E. and G. Thodos, Ind. Eng. Chem., 44, 1649 (1952).
58. Prengle, H. W., Jr., L. R. Greenhaus and R. York, Jr., Chem. Eng. Progr., 44, 863 (1048)
59. Goodwin, R. D., National Bureau of Standards Technical Note 653, Boulder, CO (1974).
60. Goodwin, R. D. and W. M. Haynes, National Bureau of Standards Monograph 170, Boulder, CO (1982).
61. Goodwin, R. D. and W. M. Haynes, National Bureau of Standards Monograph 169, Boulder, CO (1982).
62. Lambert, J. D., K. J. Cotton, M. W. Pailthorpe, A. M. Robinson, J. Scrivins, W. R. F. Vale and R. M. Young, Proc. Roy. Soc. (London), A231, 280 (1955).
63. Holland, P. M., H. J. M. Hanley, K. E. Grubbins and J. M. Haile, J. Phys. Chem. Ref. Data, 8, 59 (1979).
64. Diller, D. E., J. Chem. Eng. Data, 27, 240 (1982).
65. Hanley, H. J. M., W. M. Haynes and R. D. McCarty, J. Phys. Chem. Ref. Data, 6, 597 (1977).
66. Vilcu, R. and A. Ciochina, Rev. Roum. Chim., 17, 1679 (1972).
67. LeNeindre, B., Int. J. Heat Mass Transfer, 15, 1 (1972).
68. Carmichal. L. T., J. Jacobs and B. H. Sage, J. Chem. Eng. Data, 13, 40 (1968).
69. Carmichal. L. T. and B. H. Sage, J. Chem. Eng. Data, 9, 511 (1964).
70. Carmichal. L. T., V. Berry and B. H. Sage, J. Chem. Eng. Data, 8, 281 (1963).
71. Parkinson, C. and P. Gray, J. Chem. Soc. Faraday Trans. I, 68, 1065 (1972).
72. Hanley, H. J. M., K. E. Grubbins and S. Murad, J. Phys. Chem. Ref. Data, 6, 1167 (1977).
73. Brykov, V. P., Inzh.-Fiz. Zh., 18, 82 (1970).
74. Leng, D. E., Ind. Eng. Chem., 49, 2042 (1957).
75. Michels, A. and T. Wassenaar, Physica, 16, 221 (1950).
76. Sage, B. H. and W. N. Lacey, Monograph on API Res. Proj. 37, American Petroleum Institute, New York, NY (1955).
77. Calado, J. C. G., P. Clancy, A. Heintz and W. E. Streett, J. Chem. Eng. Data, 27, 376 (1982).
78. Kolomiets, A. Y., Sov. Res., 6(6), 42 (1974).
79. Naziev, Y. M. and A. A. Abasov, Chemtech., 20(12), 756 (1968).
80. Pachaiyappan, V., S. H. Ibrahim and N. R. Kuloor, Chem. Eng., 74(4), 140 (1967).
81. Swift, G. W. and A. Migliori, J. Chem. Eng. Data, 29, 59 (1984).
82. Parkinson, C., P. Mukhapadhyay and P. Gray, J. Chem. Soc. Faraday Trans. I, 68(6), 1077 (1972).
83. Schlinger, W. G. and B. H. Sage, Ind. Eng. Chem., 44, 2454 (1952).
84. Ambrose, D. and R. Townsend, Trans. Faraday Soc., 60, 1025 (1964).
85. Mukhopadhyay, P. and A. K. Barua, Trans. Faraday Soc., 63, 2379 (1967).
86. Nohra, S. P., T. L. Kang, K. A. Kobe and J. J. McKetta, J. Chem. Eng. Data, 7, 150 (1962).
87. Van Hook, W. A., J. Chem. Phys., 46, 1909 (1967).
88. Tarakad, R. R. and R. P. Danner, AIChE J., 23, 944 (1977).
89. Tarakad, R. R. and R. P. Danner, AIChE J., 23, 685 (1977).
90. Aston, J. G., S. V. R. Mastrangelo and G. W. Moessen, J. Am. Chem. Soc., 72, 5287 (1950).

91. Chapman, S. and T. G. Cowling, The Mathematical Theory of Non Uniform Gases, Cambridge, England (1952).
92. Stiel, L. T. and G. Thodos, J. Chem. Eng. Data, 7, 234 (1962).
93. Misic, D. and G. Thodos, AIChE J., 7, 264 (1961).
94. Scott, R. B., C. H. Meyers, R. D. Rands, F. G. Brickwedde and N. Bekkedahl, J. Res. Natl. Bur. Std., A 35, 39 (1945).
95. Aston, J. G. and G. J. Szasz, J. Am. Chem. Soc., 69, 3108 (1947).
96. Senftleben, H., Z. Agnew Physik, 17(2), 86 (1964).
97. Mansoorian, H., K. R. Hall, J. C. Holste and P. T. Eubank, J. Chem. Thermo., 13, 1001 (1981).
98. Hsu, C. C. and J. J. McKetta, J. Chem. Eng. Data, 9(1), 45 (1964).
99. Campbell, A. N. and R. M. Chatterjee, Can. J. Chem., 46, 575 (1968).
100. Majer, V., L. Svab and V. Svoboda, J. Chem. Thermo., 12, 843 (1980).
101. Campbell, A. N. and R. M. Chatterjee, Can. J. Chem., 47, 3893 (1969).
102. Harrison, R. D. and M. Hughes, Proc. Royal Soc. (London), 239a, 230 (1957).
103. Rastorguev, Y. L. and Y. A. Ganiev, Izv. Vyssh. Uchebn. Zaved. Neft. Gaz., 10(1), 79 (1967).
104. Paniego, A. R., J. A. B. Lluna and J. E. H. Garcia, Am. Quim., 71, 349 (1975).
105. VanVelzen, D., Lopes, R. Cardozo and H. Langenkamp, EUR 4735e, Commission of European Communities, Luxembourg (1972).
106. Rutherford, W. M., J. Chem. Eng. Data, 29, 163 (1984).
107. Archer, W. L. and V. L. Stevens, Ind. Eng. Chem., Prod. Res. Dev., 16(4), 319 (1977).
108. Phillips, T. W. and K. P. Murphy, J. Chem. Eng. Data, 15(2), 304 (1970).
109. Shelton, L. G., D. E. Hamilton and R. H. Fisackerly, VINYL AND DIENE MONOMERS, Part 3, E. Leonard (editor), pages 1205-1289, Wiley-Science, New York, NY (1971).
110. Reichenberg, D., (a) DSC Rep. 11, National Physical Laboratory, Teddington, England (August, 1971); (b) AIChE J., 19, 854 (1973) and (c) AIChE J., 21, 181 (1975).
111. Stiel, L. T. and G. Thodos, J. Chem. Eng. Data, 10, 266 (1964).
112. Mumford, S. A. and J. W. C. Phillips, J. Chem. Soc., 75 (1950).
113. Roy, D. and G. Thodos, Ind. Eng. Chem. Fundam., 7, 529 (1968).
114. Roy, D. and G. Thodos, Ind. Eng. Chem. Fundam., 9, 71 (1970).
115. Riedel, L., Chem. Ingr. Tech., 23(13), 321 (1951).
116. Venart, J. E. S., J. Scient. Instrum., 41(12), 727 (1964).
117. Mason, H. L., Trans. Amer. Soc. Mech. Engrs., 76(5), 817 (1954).
118. Mallan, G. M., "Thermal Conductivity of Liquids", Ph. D. Thesis, Univ. of Southern Calif., Los Angeles, CA (1968).
119. Cox, J. D. and G. Pitcher, THERMOCHEMISTRY OF ORGANIC AND ORGANOMETALLIC COMPOUNDS, Academic Press, New York, NY (1970).
120. Gordon, J. and W. F. Giauque, J. Am. Chem. Soc., 70, 1506 (1948).
121. Wagman, D. D., W. H. Evans, V. B. Parker, I. Harlow, S. M. Bailey and R. H. Schumm, Nat. Bur. Stand. Tech. Note 270-3, Washington, DC (1968).
122. Lee, B. I. and M. G. Kesler, AIChE J., 21(3), 510 (1975).
123. Paniego, A. R. and J. M. G, Pinto, Anal. de Fisca, 64(11), 343 (1969).
124. Vines, R. G. and L. A. Bennett, J. Chem. Phys., 22(3), 360 (1954).
125. Jamieson, D. T., J. B. Irving and J. S. Tudhope, LIQUID THERMAL CONDUCTIVITY - A DATA SURVEY TO 1973, Crown Publishing, Edinburg, England (1975).
126. Wilhoit, R. C. and B. J. Zwolinski, PHYSICAL AND THERMODYNAMIC PROPERTIES OD ALIPHATIC ALCOHOLS, J. Phys. Chem. Ref. Data, 2, Suppl. No. 1 (1973).
127. Ambrose, D. and C. H. S. Sprake, J. Chem. Thermo., 2, 631 (1970).
128. Kubicek, A. J. and P. T. Eubank, J. Chem. Eng. Data, 17(2), 233 (1972).
129. Kemme, H. R. and S. I. Kreps, J. Chem. Soc., 3697 (1965).
130. Won, Y. S., D. K. Chung and A. F. Mills, J. Chem. Eng., 26(2), 140 (1981).

131. Hales, J. L. and J. H. Ellender, J. Chem. Thermo., <u>8</u>, 1177 (1976).
132. Machado, J. R. S. and W. B. Streett, J. Chem. Eng. Data, <u>28</u>, 218 (1983).
133. Reid, R. C. and L. Belenyessy, J. Chem. Eng. Data, <u>5(2)</u>, 150 (1960).
134. Tarzimanov, A. A. and V. E. Mashirov, THERMOPHYSICAL PROPERTIES OF MATTER AND SUBSTANCES, <u>2</u>, V. A. Rabinovich (editor), Amerind Pub. Co., Ltd., New Delhi, India (1974).
135. Frurip, D. J., L. A. Curtiss and M. Blander, Int. J. Thermophysics, <u>2(2)</u>, 115 (1981).
136. Raal, J. D. and R. L. Rijsdijk, J. Chem. Eng. Data, <u>26</u>, 351 (1981).
137. Venart, J. E. S. and C. Krishnamurthy, National Bureau Standards Special Publication 302 (1967).
138. Riedel, L., Chem. Ing. Tech., <u>26</u>, 83 (1954).
139. Pedley, J. B. and J. Rylance, "Sussex - N. P. L. Computer Analysed Thermochemical Data", University of Sussex, Brighton, England (1977).
140. Mansson, M., P. Sellers, G. Stridh and S. Sunner, J. Chem. Thermo., <u>9</u>, 91 (1977).
141. Counsell, J. F., E. B. Lee and J. F. Martin, J. Chem. Soc. (A), 1819 (1968).
142. Hutchinson, E. and L. G. Barley, Z. Phys. Chem. Neue Folge, <u>21</u>, 30 (1959).
143. Hovorka, F., H. P. Lankelina and S. C. Stanford, J. Am. Chem. Soc., <u>60</u>, 820 (1938).
144. Jobst, W., J. Heat Mass Transfer, <u>7</u>, 725 (1964).
145. Muller, A., Fette U. Sefen, <u>49</u>, 572 (1942).
146. Hovorka, F., J. Am. Chem. Soc., <u>60</u>, 820 (1938).
147. Ambrose, D. and R. Townsend, J. Chem. Soc., 3614 (1963).
148. Brazknikov, M. M., et. al., J. Appl. Chem. USSR, <u>48(10)</u>, 2181 (1975).
149. Newsham, D. M. T. and E. J. Mendex-Lecanda, J. Chem. Thermo., <u>14</u>, 291 (1982).
150. Krone, L. H. and R. C. Johnson, AIChE J., <u>2</u>, 552 (1956).
151. Ambrose, D., J. F. Counsell, I. J. Laurenson and G. B. Lewis, J. Chem. Thermo., <u>10</u>, 1033 (1978).
152. Hoffman, S. P., J. L. San Jose and R. C. Reid, M.I.T. Industrial Liaison Program, Dept. of Chem. Eng., M. I. T., Cambridge, MA.
153. Oetting, F. L., J. Phys. Chem., <u>67</u>, 2757 (1963).
154. Andon, R., J. Counsell and F. Martin, Trans. Faraday Soc., <u>59</u>, 1555 (1963).
155. Andon, R. J. L., J. E. Connett, J. F. Counsell, E. B. Lees and J. F. Martin, J. Chem. Soc. A, 661 (1971).
156. Paz Asdrade, M. I., J. M. Paz and E. Rechacho, An. Quim., <u>66</u>, 961 (1970).
157. Swietoslawski, W. and A. Zielenkiewicz, Bull. Acad. Pol. Sci. Ser. Sci. Chim., <u>8</u>, 651 (1960).
158. Pal, A. K., Indian J. Phys., <u>41</u>, 823 (1967).
159. Rezk, H. A. and I. M. Elanivar, Z. Phys. Chem. (Lepzig), <u>245(5-6)</u>, 299 (1970).
160. Janelli, L., A. Rakshit and A. Sacco, Z. Naturforsch A, <u>29</u>, 355 (1974).
161. Poltz, H. and R. Jugel, Int. J. Heat Mass Transfer, <u>10(8)</u>, 1075 (1967).
162. Sakiadis, B. C. and J. Coates, AIChE J., <u>3</u>, 275 (1955).
163. Mallan, G. M., M. S. Michaelian and F. J. Lockhart, J. Chem. Eng. Data, <u>17(4)</u>, 412 (1972).
164. Scheffy, W. J. and E. F. Johnson, J. Chem. Eng. Data, <u>6(2)</u>, 245 (1961).
165. Curme, G. O., GLYCOLS, Reinhold Publishing Corp., New York, NY (1953).
166. Dow Chemical Company, "Organic Chemicals/The Glycols", Midland, MI (1973).
167. Union Cabide Chemical Company, "Glycols", product bulletin.
168. Ambrose, D. and D. J. Hall, J. Chem. Thermo., <u>13</u>, 61 (1981).
169. Benson, S., THERMOCHEMICAL KINETICS, 2nd edition, John Wiley & Sons, New York, NY (1976).
170. Stephens, M. A. and W. S. Tamplin, J. Chem. Eng. Data, <u>24</u>, 81 (1979).
171. Union Carbide Chemical Company, "Tables of Physical Properties", 17th edition (1960).
172. Rastorguev, Y. L. and M. A. Gazdiev, Inzh-fiz. Zh., <u>17(1)</u>, 72 (1969).
173. Ganiev, Yu. A., Zh. fiz. Khim., <u>43(1)</u>, 239 (1969).
174. Rastorguev, Y. L. and M. A. Gazdiev, Zh. fiz. Khim., <u>45(3)</u>, 692 (1971).
175. Vanderkooi, W. N., D. L. Hilderbrand and D. R. Stull, J. Chem. Eng. Data, <u>12(3)</u>, 377 (1967).
176. Slawecki, T. K. and M. C. Molstad, Ind. Eng. Chem., <u>48(6)</u>, 1100 (1956).
177. Smith, J. M. and H. C. Van Ness, INTRODUCTION TO CHEMICAL ENGINEERING

THERMODYNAMICS, 4th edition, McGraw-Hill, New York, NY (1987).
178. American Petroleum Research Project 62, Bartlesville Energy Research Center, Report No. 12, Bartlesville, OK (1971).
179. Hossenlopp, I. A. and D. W. Scott, J. Chem. Thermo., 13, 415 (1981).
180. Weber, J. H., AIChE J., 2, 514 (1956).
181. Nichols, W. B., H. H. Reamer and B. H. Sage, Ind. Eng. Chem., 47, 2219 (1955).
182. Cook, M. W., Rev. Sci. Instr., 29, 399 (1958).
183. McMiking, J. H., "Vapor Pressures and Saturated Liquid and Vapor Densities of Isomeric Heptanes and Octanes", Ph. D. Thesis, Ohio State University, Columbus, OH (1961).
184. Douglas, T. B., G. T. Furikawa, R. E. McCoskey and A. F. Ball, J. Res. Natl. Bur. Std., 53(3), 139 (1954).
185. Messerly, J. F., G. B. Gutherie, S. S. Todd and H. L. Finke, J. Chem. Eng. Data, 12, 338 (1967).
186. Gugor'ev, H. A., Y. L. Rastorguev and G. S. Yanin, Iz. Vyssh. Uchebn. Zaved. Neft. Gaz., 18(10), 63 (1975).
187. Conolly, T. J., B. H. Sage and W. N. Lacey, Ind. Eng. Chem. 43, 946 (1951).
188. Smith, W. J. S., L. D. Durbin and R. Kobayashi, J. Chem. Eng. Data, 5, 316 (1960).
189. McKelvey, F. E., Hydrocarbon Processing, 43(7), 146 (1964).
190. McCoubrey, J. C. and N. M. Singh, J. Phys. Chem., 67, 517 (1963).
191. Geist, G. M. and M. R. Cannon, Ind. Eng. Chem., Anal. Ed., 18, 611 (1946).
192. Carmichael, L. T., J. Jacobs and B. H. Sage, J. Chem. Eng. Data, 14(1), 31 (1969).
193. Vines, R. G., Australian J. Chem., 6, 1 (1953).
194. Lambert, J. D., E. N. Staines and S. D. Woods, Proc. Royal Soc. (London), A200, 262 (1950).
195. Mustafaev, R. A., Teplofiz. Vys. Temp., 12(4), 883 (1974).
196. Tarzimanov, A. A. and V. E. Mashirov, Teploenergetika, 14(12), 67 (1967).
197. Finke, H. L., M. E. Gross, G. Waddington and H. M. Huffman, J. Am. Chem. Soc., 78, 854 (1954).
198. Nagasaka, Y. and A. Nagashima, Ind. Eng. Chem. Fundam., 20, 216 (1981).
199. Rastorguev, Y. L., G. F. Bogatov and B. A. Grigor'ev, Iz. Vyssh. Ucheb. Zaved. Neft. Gaz., 11(12), 59 (1968).
200. Miner, C. S., GLYCEROL, Reinhold Publishing Corp., New York, NY (1953).
201. Soap and Detergent Association, "Physical Properties of Glycerine and Its Solutions", New York, NY.
202. Glycerine Producers Association, "Glycerine - Properties, Reactions and Performance", New York, NY.
203. Dow Chemical Company, "Glycols: Properties and Uses", product bulletin, Midland, MI.
204. Union Carbide Chemical Company, "Glycols", product bulletin, New York, NY.
205. Omelchenko, F. D., Iz. Vyssh. Ucheb. Zayed. Pisch. Tekh., 3, 97 (1962).
206. Costello, J. M. and S. T. Bowden, Recueil, 77, 36 (1958).
207. Segur, J. B. and H. Oberstar, Ind. Eng. Chem., 43, 2117 (1951).
208. Bates, O. K. and G. Hazzard, Ind. Eng. Chem., 37, 193 (1945).
209. Woolf, J. and W. Sibbert, Ind. Eng. Chem., 46, 1947 (1954).
210. Scheffy, W. J. and E. F. Johnson, J. Chem. Eng. Data, 6, 245 (1961).
211. Ross, G. R. and W. J. Heideger, J. Chem. Eng. Data, 7(4), 505 (1962).
212. Othmer, D. F. and E. Yu, Ind. Eng. Chem., 60, 22 (1968).
213. Nakanishi, K., T. Matsumoto and H. Mitsuyoshi, J. Chem. Eng. Data, 16, 44 (1971).
214. Riedel, L., Chem. Ingr. Tech., 23(19), 465 (1951).
215. Vargaftik, N. B., Igv. Vses. Teplotekhn. Inst., 20 (1951).
216. Challoner, A. R. and R. W. Powell, Proc. Roy. Soc. A, 238, 1212, 90 (1956).
217. Rastorguev, Y. L. and Y. A. Ganiev, Inzhfiz. Zh., 14(4), 689, 697 (1968).
218. Filippov, L. P., Vestnik. Mosk. gos. Univ. Ser. 3 Fiz. Astron., 15(3), 61 (1960).
219. Camin, D. L. and F. D. Rossini, J. Phy. Chem., 60, 1446 (1956).
220. Wright, F. J., J. Chem. Eng. Data, 6(3), 454 (1961).
221. Forziati, A. D., D. L. Camin and F. D. Rossini, J. Res. Natl. Bur. Std., 45, 406 (1950).

222. Naziev, Y. M. and A. A. Abasov, Inter. Chem. Eng., 10(2), 270 (1970).
223. Wolfe, D. B., M. S. Thesis, Ohio State University, Columbus, OH (1970).
224. Wright, F. J., J. Chem. Eng. Data, 6, 454 (1961).
225. Naziev, Y. M. and A. A. Abasov, Khim. Tekhnal. Topl. Masel., 15(3), 22 (1970).
226. Naziev, Y. M. and A. A. Abasov, Iz. Vyssh. Uchebn. Zaved. Neft. Gaz., 12(1), 81 (1969).
227. Das, T. R., C. O. Reed and P. T. Eubank, J. Chem. Eng. Data, 22(1), 9 (1977).
228. Isaac, R., K. Li and L. U. Canjar, Ind. Eng. Chem., 46, 199 (1954).
229. Morecroft, D. W., J. Inst. Petrol., 44, 433 (1958).
230. Goodwin, R. D., Nat. Bur. Std., NBSIR 79-1612, Thermophysical Properties Division, National Engineering Laboratory, Boulder, CO (July, 1979).
231. Sugisaki, M., K. Adachi, S. Hirochi and S. Saki, Bull. Chem. Soc. Japan, 41, 593 (1968).
232. Messerly, J. F. and H. L. Finke, J. Chem. Thermo., 3, 675 (1971).
233. Vilim, O., Collection Czech. Chem. Comm., 25, 993 (1960).
234. Ryabtsev, N. I. and V. A. Kazaryan, Gazov. Delo., 2, 36 (1970).
235. Hess, L. G. and V. V. Tilton, Ind. Eng. Chem., 42(6), 1251 (1950).
236. Bott, T. R. and H. N. Sadler, J. Chem. Eng. Data, 11, 25 (1966).
237. McDonald, R. A., S. H. Shader and D. P. Stull, J. Chem. Eng. Data, 4(4), 311 (1959).
238. Dow Chemical Company, "Epichlorohydrin", No. 298-680-80, Midland, MI (1980).
239. Walters, C. and J. Smith, J. Chem. Eng. Data, 8(3), 281 (1952).
240. Oetling, F. L., J. Chem. Phys., 41, 149 (1964).
241. Eucken, A. Z. Physik., 14, 324 (1913).
242. Dow Chemical Company, "Ethylene Oxide and Propylene Oxide - Product Bulletin", Midland, MI
243. Giauque, W. F. and J. Gordon, J. Am. Chem. Soc., 71, 2176 (1949).
244. Mock, J. E. and J. M. Smith, Ind. Eng. Chem., 42, 2125 (1950).
245. Shell Chemical Company, "Epichlorohydrin", Product Bulletin.
246. Urbancova, L., Chem. Zvesti., 13, 224 (1959).
247. Sinke, G. C., J. Chem. Eng. Data, 7, 74 (1962).
248. Union Carbide Chemicals Company, "Alkylene Oxides:, F-40558, New York, NY (1961).
249. Dow Chemical Company, "Alkylene Oxides from Dow", No. 110-551-69R, Midland, MI (1969).
250. Huffman, H. M., M. E. Gross, D. W. Scott and J. P. McCullough, J. Phys. Chem., 65, 495 (1961).
251. Kay, W. B. and F. M. Warzel, Ind. Eng. Chem., 43, 1150 (1951).
252. Svoboda, V., V. Charvatova, V. Majer and V. Hynek, Coll. Czech. Chem. Comm., 47(2), 543 (1981).
253. McMicking, J. H. and W. B. Kay, Proc. Am. Petrol. Inst., 45(3) (1965).
254. Eicher, L. D. and B. J. Zwolinski, J. Phys. Chem., 76, 3295 (1972).
255. Auerbach, C. E., B. H. Sage and W. N. Lacey, Ind. Eng. Chem., 42, 110 (1950).
256. Landolt-Bornstein, ZAHLENWERTE UND FUNKIONEN ANS PHYSIK, CHEMEI, ASTRONOMIE UND TECHNIK, 6th ed., Springer-Verlag, Berlin, Germany (1972).
257. Kerimov, A. M., F. G. El'Darov and V. S. El'Darov, Iz. Vyssh. Uchebn. Zaved. Neft. Gaz., 13(1), 77 (1970).
258. Kay, W. B., J. Am. Chem. Soc., 68, 1336 (1946).
259. Ambrose, D., et. al., Trans. Faraday Soc., 56, 1452 (1960).
260. Beattie, J. A. and D. G. Edwards, J. Am. Chem. Soc., 70, 3382 (1948).
261. Ambrose, D., C. H. S. Sprake and R. Townsend, J. Chem. Soc., Faraday Trans., 69, 839 (1973).
262. Chao, J., A. S. Rodgers, R. C. Wilhoit and B. J. Zwolinski, J. Phys. Chem, Ref. Data, 3(1), 141 (1974).
263. Andon, R. J. L., J. F. Counsell, D. A. Lee and J. F. Martin, J. Chem. Soc., Faraday Trans., 69, 1721 (1973).
264. E. I. Du Pont de Nemours & Co., "Thermodynamic Properties of Freon 12", Wilmington, DE (1956).
265. E. I. Du Pont de Nemours & Co., "Thermodynamic Properties of Freon 11", Wilmington, DE (1965).
266. Heckle, M., Chem. Ingr. Techn., 41(3), 757 (1969).
267. Backstrom, M., Kylte. Tidksr., 24(2), 40 (1965).

268. Tsvetkov, O. B., Inzh-fiz. Zh., 9(1), 42 (1965).
269. Powell, R. W., Proc. 10th Internatl. Congress Refrig., 1, 382 (1959).
270. Cherneeva, L., Kholod. Tekh., 29(3), 55 (1952).
271. Danilova, G., Kholod. Tekh., 28(2), 22 (1951).
272. Widmer, F., Kaltetechnik, 14(2), 38 (1962).
273. Kobe, K. A., Petroleum Refiner, 30(11), 151 (1951).
274. Rubin, T. R., et. al., J. Am. Chem. Soc., 66, 279 (1944).
275. Mason, H. L., Trans. ASME, 817 (July, 1954).
276. Altunin, V. V., V. Z. Geller, E. K. Petrov, D. C. Rasskazov, and G. A. Spiridonov, THERMOPHYSICAL PROPERTIES OF FREONS, Methane Series, Part 1, Hemisphere Publishing Corporation, New York, NY (1987).
277. Altunin, V. V., V. Z. Geller, E. A. Kremenevskaya, I. I. Perelshtein, and E. K. Petrov, THERMOPHYSICAL PROPERTIES OF FREONS, Methane Series, Part 2, Hemisphere Publishing Corporation, New York, NY (1987).
278. E. I. Du Pont de Nemours & Co., "Freon Fluorocarbons", Technical Bulletin B-2, Wilmington, DE.
279. Am. Soc. Refrig. Engrs., "Air Conditioning Refrigeration Data Book", New York, NY (1955).
280. Booth, H. S. and C. F. Swinehart, J. Am. Chem. Soc., 57, 1337 (1935).
281. Hovorka, F. and F. E. Geiger, J. Am. Chem. Soc., 55, 4759 (1933).
282. Benning, A. F. and R. C. McHarness. Ind. Eng. Chem., 32, 497 (1940).
283. Martin, J. J., J. Chem. Eng. Data, 5(3), 334 (1960).
284. Gelles, E. and K. S. Pitzer, J. Am. Chem. Soc., 75, 5259 (1953).
285. Benning, A. F. and R. C. McHarness. Ind. Eng. Chem., 32, 976 (1940).
286. Neilson, E. F. and D. White, J. Am. Chem. Soc., 79, 5618 (1957).
287. Benning, A. F. and R. C. McHarness. Ind. Eng. Chem., 32, 814 (1940).
288. Wizell, O. W. and J. W. Johnson, Technical Paper, Am. Soc. Heat.-Refrig.-Air Cond. Engineers, Semi-annual Meeting (January, 1965).
289. Makita, T., Rev. Phy. Chem. (Japan), 24, 74 (1954).
290. Hou, Y. C. and J. J. Martin, AIChE J., 5, 125 (1961).
291. Mears, W. H., et. al., Ind. Eng. Chem., 47, 1449 (1955).
292. Valentine, R. H., et. al., J. Phy. Chem., 66, 392 (1962).
293. Pace. E. L. and J. C. Aston, J. Am. Chem. Soc., 70, 566 (1948).
294. Wicklund, J. S., et. al., J. Res. Nat. Bur. Stands., 51, 91 (1953).
295. Dreisbach, R. R. and S. A. Shrader, Ind. Eng. Chem., 41, 2879 (1949).
296. Missenard, F. A., Revue Generale Thermique, 5(50), 125 (1966).
297. Ketelaar, J. and N. Van Meurs, Rec. Trav. Chim., 76, 437 (1957).
298. Gladstone, J. H., J. Chem. Soc., 45, 241 (1884).
299. Biron, E. B., Z. Phys. Chem., 81, 590 (1912).
300. Brucksch, W. F. and W. T. Ziegler, J. Chem. Phys., 10, 740 (1942).
301. Gelles, E. and K. S. Pitzer, J. Am. Chem. Soc., 75, 5259 (1953).
302. Pitzer, K. S., J. Am. Chem. Soc., 62, 331 (1940).
303. Heston, W. M., E. J. Hennely and C. P. Smyth, J. Am. Chem. Soc., 72, 2071 (1950).
304. Baroncini, C., F. DiFilippo, G. Latini and M. Pacetti, Internat. J. Thermophysics, 2(1), 21 (1981).
305. Stull, D. B., Ind. Eng. Chem., 39, 517 (1947).
306. Egan, C. J. and J. D. Kemp, J. Am. Chem. Soc., 60, 2097 (1938).
307. Railing, W. E., J. Am. Chem. Soc., 61, 3349 (1939).
308. Reed, J. F. and B. S. Rabinovitch, J. Chem. Eng. Data, 2(1), 75 (1957).
309. Kimura, O., J. Chem. Soc. (Japan), 63, 814 (1942).
310. Masia, A. P. and M. D. Alvarez, Dept. of Commerce, Office of Technical Service, PB Report 147,614, Washington, DC (1961).
311. Venart, J. E. S., J. Chem. Eng. Data, 10(3), 239 (1965).

Appendix A

CONVERSION TABLES

1. **Temperature**
   To convert from Centigrade to:
   Kelvin, add 273.15
   Rankine, multiply Kelvin by 1.8
   Fahrenheit, multiply Centigrade by 1.8 and add 32

2. **Pressure**
   To convert from psia to:
   kPa, multiply by 6.895
   psig, subtract 14.7
   mm Hg, multiply by 51.7
   atmospheres, divide by 14.7
   bars, divide by 14.89

3. **Heat of Vaporization**
   To convert from kJ/kg to:
   BTU/lb, multiply by 0.43
   cal/gram, multiply by 0.239

4. **Density**
   To convert from g/ml to:
   lb/ft^3, multiply by 62.43
   lb/gallon, multiply by 8.345

5. **Surface Tension**
   To convert from dynes/cm to:
   N/m, multiply by 0.001

6. **Heat Capacity**
   To convert from J/g K to:
   BTU/lb R, multiply by 0.239
   cal/gram K, multiply by 0.239

7. **Viscosity**
   To convert from micropoise to:
   lb/ft s, multiply by 0.0672E-06
   centipoise, multiply by 1.0E-04
   poise, multiply by 1.0E-06
   Pa s (pascal seconds), multiply by 1.0E-07

   To convert from centipoise to:
   lb/ft s, multiply by 0.000672
   micropoise, multiply by 10,000
   poise, multiply by 0.01
   Pa s (pascal seconds), multiply by 0.001

8. **Thermal Conductivity**
   To convert from W/m K to:
   BTU/hr ft R, multiply by 0.5788
   calorie/cm s K, multiply by 0.002388

9. **Enthalpy of Formation**
   To convert from kJ/mol to:
   kcal/mol, multiply by 0.239

10. **Gibb's Free Energy of Formation**
    To convert from kJ/mol to:
    kcal/mol, multiply by 0.239

Appendix B

COMPOUND INDEX BY FORMULA

Formula	Name	Synonym	Page
CBrF3	bromotrifluoromethane		235
CCl4	carbon tetrachloride		44
CClF3	chlorotrifluoromethane	fluorocarbon-13	205
CCl2F2	dichlorodifluoromethane	fluorocarbon-12	205
CCl3F	trichlorofluoromethane	fluorocarbon-11	205
CF4	tetrafluoromethane	fluorocarbon-14	205
CHCl3	chloroform	trichloromethane	44
CHClF2	chlorodifluoromethane	fluorocarbon-21	215
CHCl2F	dichlorofluoromethane	fluorocarbon-22	215
CHF3	trifluoromethane	fluorocarbon-23	225
CH2Cl2	methylene chloride	dichloromethane	44
CH3Br	methyl bromide		235
CH3CL	methyl chloride		44
CH4	methane		1
CH4O	methanol		74
C2Cl4	perchloroethylene		54
C2Cl2F4	1,2-dichloro-1,1,2,2-tertafluoroethane	fluorocarbon-114	215
C2Cl3F3	1,1,2-trichloro-1,2,2-trifluoroethane	fluorocarbon-113	215
C2F6	hexafluoroethane	fluorocarbon-116	225
C2HCl3	trichloroethylene		54
C2H2	acetylene	ethyne	24
C2H2Cl2	vinylidene chloride		54
C2H2Cl2	cis-1,2-dichloroethylene		195
C2H2Cl2	trans-1,2-dichloroethylene		195
C2H2Cl4	1,1,2,2-tertachloroethane		195
C2H2F2	vinylidene fluoride		225
C2H3Cl	vinyl chloride		54
C2H3Cl3	methyl chloroform	1,1,1-trichloroethane	195
C2H3F	vinyl fluoride		225
C2H2	acetylene	ethyne	24
C2H4	ethylene	ethene	14
C2H4Br2	ethylene dibromide	1,2-dibromoethane	235
C2H4Cl2	ethylene dichloride	1,2-dichloroethane	64
C2H4O	ethylene oxide		104
C2H5Br	ethyl bromide	bromoethane	235
C2H5Cl	ethyl chloride	chloroethane	64
C2H6	ethane		1
C2H6O	ethanol	ethyl alcohol	74
C2H6O2	ethylene glycol		114
C3H4	methyl acetylene	propyne	24

C3H4	propadiene	allene	34
C3H5ClO	epichlorohydrin		104
C3H6	propylene	propene	14
C3H6Cl2	propylene dichloride	1,2-dichloropropane	64
C3H6O	propylene oxide		104
C3H6O	allyl alcohol		94
C3H7Cl	propyl chloride	1-chloropropane	64
C3H8	propane		1
C3H8O	propanol	propyl alcohol	74
C3H8O	isopropanol	isopropyl alcohol	84
C3H8O2	propylene glycol	1,2-propylene glycol	135
C3H8O3	glycerine	glycerol	135
C4H6	1-butyne		24
C4H6	2-butyne		24
C4H6	1,2-butadiene		34
C4H6	1,3-butadiene		34
C4H8	1-butene		14
C4H8	trans 2-butene		14
C4H8	cis 2-butene		14
C4H8	isobutylene	2-methylpropene	175
C4H8O	butylene oxide	1,2-butylene oxide	104
C4H10	butane		1
C4H10	isobutane	2-methylpropane	175
C4H10O	butanol	butyl alcohol	74
C4H10O	isobutyl alcohol		84
C4H10O	sec-butyl alcohol		84
C4H10O	tert-butyl alcohol		84
C4H10O3	diethylene glycol		114
C5H8	isoprene	2-methyl-1,3-butadiene	175
C5H10	1-pentene	isopentene	165
C5H12	pentane		155
C5H12	isopentane	2-methylbutane	175
C5H12O	1-pentanol	1-amyl alcohol	94
C6H12	1-hexene		165
C6H14	hexane		155
C6H14	2-methylpentane	isohexane	185
C6H14O	1-hexanol	1-hexyl alcohol	94
C6H14O3	dipropylene glycol		135
C6H14O4	triethylene gylcol		114
C7H14	1-heptene		165
C7H16	heptane		155
C7H16	2-methylhexane		185
C7H16O	1-heptanol	1-heptyl alcohol	94
C8H16	1-octene		165
C8H18	octane		155
C8H18	isooctane	2,2,4-trimethylpentane	185
C8H18	2-methylheptane		185

Appendix C

COMPOUND INDEX BY NAME

Name	Synonym	Page

```
acetylene...............ethyne..................24
allyl alcohol.......................................94
bromotrifluoromethane..............................235
1,2-butadiene......................................34
1,3-butadiene......................................34
butane..............................................1
butanol.................butyl alcohol..............74
1-butene..........................................14
trans 2-butene....................................14
cis 2-butene......................................14
sec-butyl alcohol.................................84
tert-butyl alcohol................................84
butylene oxide..........1,2-butylene oxide......104
1-butyne..........................................24
2-butyne..........................................24
carbon tetrachloride..............................44
chlorodifluoromethane....fluorocarbon-21.........215
chloroform..............trichloromethane........44
chlorotrifluoromethane...fluorocarbon-13.........205
dichlorodifluoromethane..fluorocarbon-12.........205
cis-1,2-dichloroethylene..........................195
trans-1,2-dichloroethylene........................195
dichlorofluoromethane....fluorocarbon-22.........215
1,2-dichloro-1,1,
  2,2-tertafluoroethane...fluorocarbon-114.......215
diethylene glycol................................114
dipropylene glycol...............................135
epichlorohydrin..................................104
ethane.............................................1
ethanol.................ethyl alcohol..............74
ethyl bromide...........bromoethane.............235
ethyl chloride..........chloroethane............64
ethylene................ethene..................14
ethylene dibromide......1,2-dibromoethane.......235
ethylene dichloride.....1,2-dichloroethane......64
ethylene glycol..................................114
ethylene oxide...................................104
glycerine...............glycerol................135
heptane..........................................155
1-heptanol..............1-heptyl alcohol........94
1-heptene........................................165
hexafluoroethane........fluorocarbon-116........225
```

hexane		155
1-hexanol	1-hexyl alcohol	94
1-hexene		165
isobutane	2-methylpropane	175
isobutyl alcohol		84
isobutylene	2-methylpropene	175
isooctane	2,2,4-trimethylpentane	185
isopentane	2-methylbutane	175
isoprene	2-methyl-1,3-butadiene	175
isopropanol	isopropyl alcohol	84
methane		1
methanol		74
methyl acetylene	propyne	24
methyl bromide		235
methyl chloride		44
methyl chloroform	1,1,1-trichloroethane	195
methylene chloride	dichloromethane	44
2-methylheptane		185
2-methylhexane		185
2-methylpentane	isohexane	185
octane		155
1-octene		165
pentane		155
1-pentanol	1-amyl alcohol	94
1-pentene	isopentene	165
perchloroethylene		54
propadiene	allene	34
propane		1
propanol	propyl alcohol	74
propyl chloride	1-chloropropane	64
propylene	propene	14
propylene dichloride	1,2-dichloropropane	64
propylene oxide		104
propylene glycol	1,2-propylene glycol	135
1,1,2,2-tertachloroethane		195
tetrafluoromethane	fluorocarbon-14	205
trichloroethylene		54
trichlorofluoromethane	fluorocarbon-11	205
1,1,2-trichloro-1,2,2-trifluoroethane	fluorocarbon-113	215
triethylene gylcol		114
trifluoromethane	fluorocarbon-23	225
vinyl chloride		54
vinyl fluoride		225
vinylidene chloride		54
vinylidene fluoride		225